Physical Aspects of Brachytherapy

Medical Physics Handbooks
Other books in the series

1 **Ultrasonics**
J P Woodcock

2 **Computing Principles and Techniques**
B L Vickery

3 **Physical Principles of Audiology**
P M Haughton

4 **Urodynamics**
D J Griffiths

5 **Thermoluminescence Dosimetry**
A F McKinlay

8 **Nuclear Particles in Cancer Treatment**
J F Fowler

9 **Physics and the Circulation**
J O Rowan

10 **The Evaluation of Medical Images**
A Ll Evans

11 **Ultraviolet Radiation in Medicine**
B L Diffey

12 **Radionuclide Techniques in Clinical Investigation**
P W Horton

13 **Physics of Electron Beam Therapy**
S C Klevenhagen

14 **Radiotherapy Treatment Planning** (2nd edn)
R F Mould

15 **Fundamentals of Radiation Dosimetry** (2nd edn)
J R Greening

16 **Radiation Protection**
R L Kathren

17 **Linear Accelerators for Radiation Therapy**
D Greene

18 **Non-ionising Radiation**
H Moseley

Series Editor: **Professor J M A Lenihan**
Department of Nursing Studies, University of Glasgow

Medical Physics Handbooks 19

Physical Aspects of Brachytherapy

T J Godden
Bristol Radiotherapy and Oncology Centre

Adam Hilger, Bristol and Philadelphia
in collaboration with the
Hospital Physicists' Association

British Library Cataloguing in Publication Data

Godden, T. J.
Physical aspects of brachytherapy.—
(Medical physics handbooks, ISSN 0143-0203; 19).
1. Cancer—Radiotherapy 2. Radio-isotope brachytherapy
I. Title II. Hospital Physicists' Association III. Series
616.99′40642 RC2/1.R3

ISBN 0-85274-511-/

Library of Congress Cataloging-in-Publication Data

Godden, T. J.
Physical aspects of brachytherapy.
(Medical physics handbooks; 19)
Bibliography: p.
Includes index.
1. Radioisotope brachytherapy. 2. Cancer—Radiotherapy. I. Title. II. Series.
[DNLM: 1. Brachytherapy. W1 ME409V v.19/ WN 450 G578p]
RC271.R27G63 1988 615.8′424 87–25149

ISBN 0-85274-511-7

Published under the Adam Hilger imprint by IOP Publishing Ltd
Techno House, Redcliffe Way, Bristol BS1 6NX, England
242 Cherry Street, Philadelphia, PA 19106, USA

Printed in Great Britain

Contents

Preface

Brachytherapy is the term used to describe the interstitial, intracavitary and surface application of discrete sealed sources in the treatment of malignant disease. This book has been written with the aim of providing the reader with an overall view of historical and current brachytherapy practice and sources available and to give an insight into the associated dosimetry and a practical guide to the use of dosage systems in clinical situations. It is therefore a practical teaching text for those working within this field of therapy, for junior registrars in departments of radiotherapy and oncology studying for the First Examination for the Fellowship of the Royal College of Radiologists (Radiotherapy and Oncology), for radiotherapy radiographers preparing for the qualifying and Higher examination of the College of Radiographers and also serves as an introduction to the subject for newly qualified physicists.

The first chapter of the book gives details of the historical background to brachytherapy, definitions of the units commonly used and a summary of the biological effect of continuous radiation on tissues, whilst the second chapter outlines the requirements of sources for brachytherapy, their production and construction. Chapter 3, which is more orientated towards the physicist, discusses the theory of sealed-source dosimetry, whilst measurement techniques for determining source strength and the relative dose distribution around sources are discussed in Chapter 4, together with the many ways in which source strength has been specified over the years. Many dosage systems have been developed for use in brachytherapy. The major dosage systems are reviewed and their practical use for external applicators and in interstitial therapy is detailed in Chapters 5 and 6, the emphasis being on the practical aspects of how treatments are planned and how the distribution rules and dose tables of the various dosage systems can be used. The various approaches to intracavitary brachytherapy are considered in Chapter 7, together with the different clinical techniques adopted. This chapter also includes discussion on the relative merits of different forms of after-loading techniques. Chapter 8, again, is more orientated to the physicist since it discusses radiographic techniques for source localisation and computer algorithms for dose calculation and source location, although the former will be of importance to the radiotherapist and the radiographer in their understanding of the limitations of positional accuracy from radiographs. In the final chapter the safe use of sources is reviewed—their handling, transportation, storage and custody.

Throughout this book two systems of units are deliberately used together. Since much of the clinical experience in brachytherapy has been gained using historical units, these are therefore included alongside the newer SI units. It is hoped that for many readers the book will act as the starting point as they seek to understand clinical brachytherapy and the associated dosimetry and that the list of references will be helpful to those who wish to carry out further study in this field of work.

I am grateful to the Series Editor of the Medical Physics Handbooks for giving me the opportunity to write this text. Throughout its preparation I have received encouragement from departmental colleagues and friends and colleagues who work in the field of radiotherapy physics. In addition I am pleased to acknowledge the help received from the staff at Adam Hilger, notably Mr Neville Hankins and Miss Sarah James. I also wish to record my thanks to Mrs M Lake for helping to type the manuscript and to Mr G Garland for preparing many of the illustrations. Finally I would like to record my appreciation to my wife Majorie for all the help and encouragement she has given during the preparation of the book.

T J Godden
Bristol

1 Sealed-source Therapy

1.1 Historical Background

Within three years of the discovery of radium by Marie Curie in 1898 the first patients were being treated using small quantities of the substance implanted into tumours. The idea of using radium for medical purposes came as a result of an observation made by Pierre Curie who attributed Henri Becquerel's skin erythema to the biological effect of radiation on tissue caused by the tube of radium salt Henri Becquerel carried in a waistcoat pocket. As a result of this observation, Pierre Curie offered a small radium tube to a Dr Danlos in Paris and suggested that he should use it to treat a tumour. Thus began the practice of brachytherapy.

The first interstitial application in the United States was carried out by Abbe in 1905 and in the same year the Holt Radium Institute was founded in Manchester by Fricke. In the ensuing decade various medical institutions in Europe and North America (notably the Holt Radium Institute, Manchester; The Memorial Hospital, New York; the Pasteur Laboratory of the Radium Institute of the University of Paris and the Radiumhemmet in Stockholm) began to develop the use of radium sources for interstitial and intracavitary therapy. Initially, two types of radium sources were used, sealed glass tubes containing a radium salt and flat applicators coated with radium and sealed with varnish. However, clinical experience with these sources led to radiation necrosis and it became clear that this was due, in part, to the intense beta-radiation emitted by the radium salt. Various workers tried to encapsulate the salt in steel or platinum casings to overcome this problem and in 1914 Stevenson and Joley manufactured the first radium needles. Unfortunately early attempts were not successful and it was not until 1920 that Failla eventually achieved the successful filtration of beta-rays by placing radon in capillary gold tubes instead of glass. The choice of suitable encapsulation materials was studied by other workers, notably Edith Quimby who in 1920 published a detailed experimental study of the effect of different filters on the radiation emitted by radium and radon and concluded that brass was a suitable filter and that a few millimetres of rubber would remove the secondary radiation.

As clinical experience grew in the use of radium and radon, workers began to realise the need to correlate the effects of the radiation on tissue with the amount of radium and radon used. The first step in this process was to define the quantity of radiation in terms of a standard radium source. This was made possible by Marie Curie who in 1912 prepared the first international radium standard. This consisted of 21.99 mg of radium chloride loosely contained in a glass tube. Whilst this standardised the quantity of radiation for a given amount of radium, it did not overcome the problem of defining the amount of radiation absorbed in tissue. The importance of this was realised by Failla who emphasised that radiation dose should be estimated in terms of the energy absorbed by tissue rather than the quantity of radium used, since this took no account of filtration, distribution, duration of irradiation, dose rate and the relative positions of the sources to each other and to normal healthy tissue.

Thus concepts of radiation units began to evolve in the early 1920s when attempts were made to provide more accurate methods of dose calculation. In 1921 Sievert described a method of calculating the dose from a filtered gamma-ray source (Sievert 1921), whilst a year later Quimby (1922) published tables of the intensity distribution at various distances from point, linear, circular, square and rectangular sources. As multiple sources began to be used in brachytherapy, with the consequent effect on radiation distribution, several methods for the systematic application of radium were proposed. Using Sievert's approach, Quimby calculated tables of dose values from linear sources and various combinations of equal strength linear sources (Quimby 1932). The other major advance providing quantification of doses from sealed sources came in a series of papers published by Paterson and Parker (Paterson and Parker 1934, 1938, Paterson *et al* 1936). These dosimetry systems became the basis of effective radium brachytherapy for more than 30 years by making it possible to calculate a single stated dose from the quantity of radium used and the geometrical arrangement of the implant.

Radium and radon continued to be used exclusively until the development of nuclear reactors after World War II made it possible for other radio-isotopes to be manufactured. In 1948 Myers in the US produced needles of cobalt-60 for interstitial therapy (Myers 1948). This was followed in the early 1950s by the use of other radionuclides such as tantalum-182 wire, gold-198 and iridium-192 (Sinclair 1952, Myers *et al*

1953, Henschke *et al* 1953, Henschke 1956, 1958). For intracavitary brachytherapy caesium-137 began to replace the use of radium in the early 1960s. The use of radium and radon is now decreasing and the artificially produced radionuclides such as caesium-137, gold-198, iridium-192, cobalt-60 and, more recently, iodine-125 are being used as replacements. Research is also taking place into the possibility of using neutrons and gamma-rays produced by californium-252 for brachytherapy purposes.

Developments have also occurred in recent decades in methods of calculating dose. The advent of computers now makes it possible to produce more readily isodose distributions around intracavitary and interstitial sources. These techniques are becoming increasingly used and to a certain extent are replacing the dose tables of Paterson and Parker, and Quimby, which we must acknowledge laid the foundation of the physics of brachytherapy dosimetry. A fuller review of the history of x-rays and gamma-rays in radiotherapy has been published by Mould (1980).

1.2 Radiation Units

1.2.1 Introduction

Early workers with radium soon acknowledged the need to quantify the patient's reaction to radiation in terms of the amount of radium used and the time for which it was applied. Estimates of dose were attempted as early as 1902 by Holzknecht who recognised that the colour change in a fused mixture of potassium chloride and sodium carbonate was a function of the exposure of the substance to radiation. Chemical dosimeters were also used for measuring dose in radium therapy by Sabouraud and Noire. This technique used pastilles of platinobarium cyanide and lead to the pastille unit as a means of quantifying dose. These chemical dosimeters were not, however, universally accepted; indeed, in 1904 Beclere cautioned against reliance on chemical indicators as a measurement of dose since the correlation between their dose response and the skin reactions seen in patients was not good. These early attempts to quantify dose were the basis of brachytherapy dosimetry. Over the years various workers have attempted to quantify patient dose either in terms of the quantity of the radioactive substance used, or activity, or the ionisation radiation produces in air, or exposure, or the energy absorbed by tissue, or absorbed dose, or a combination of two or all of these methods.

1.2.2 Activity

The concept of relating patient dose to the length of time for which radiation was applied to a patient was one of the early methods of estimating dose. In 1909 Turner suggested the unit of the milligram–hour, mg h, for brachytherapy. This unit was defined as the product of the weight of a pure salt of radium in milligrams and the time in hours for which the source was in contact with tissue. Whilst this was satisfactory for radium, the milligram–hour could not be readily used for the daughter product, radon, which was by then beginning to be used as an alternative to radium. A new unit was therefore suggested in 1910, the 'curie', this being defined as 'the amount of radon in equilibrium with one gramme of radium'. Initially, the weight of radium suggested for this unit varied with the radium salt used, but following the preparation of the first international radium standard, the definition of the curie was subsequently modified to be the weight of the pure radium element in the source. This definition continued in use for 20 years, after which its application was widened to include any member of the uranium series of naturally occurring radioactive elements, all of which would be in equilibrium with a mass of radium. This marked the development of the definition of activity by specifying the amount of material in terms of the number of nuclear transformations per second. In 1951 the International Commission on Radiological Units and Measurements (ICRU) redefined the quantity, and the associated unit, to cover all radioactive materials in the following terms (ICRU 1951):

> The Curie is the unit of radioactivity defined as the quantity of any radioactive nuclide in which the number of disintegrations per second is 3.700×10^{10}.

It has been conventionally assumed that 1 mg of radium has an activity of 1 mCi. However, measurements have shown that in reality 1 mg of radium has an activity of 0.989 mCi.

The term 'activity' (A) was first introduced by the ICRU in 1962 (ICRU 1962) and it was defined to be the quotient of the nuclear transformations, ΔN, which occur in the quantity of radioactive nuclide in a time Δt:

$$A = \Delta N / \Delta t$$

the special unit of activity being the curie (Ci) where $1\ \mathrm{Ci} = 3.7 \times 10^{10}\ \mathrm{s}^{-1}$ (exactly).

Further changes have occurred in the definition of activity in subsequent ICRU publications (ICRU 1968, 1971), the latest definition being (ICRU 1980)

> ... the activity A of an amount of radioactive nuclide in a particular energy state at a given time is the quotient of $\mathrm{d}N$ by $\mathrm{d}t$ where $\mathrm{d}N$ is the expected value of the number of spontaneous nuclear transitions from that energy state in the time interval $\mathrm{d}t$.

That is,

$$A = \mathrm{d}N/\mathrm{d}t$$

where the particular energy state is the ground state of the nuclide unless otherwise specified.

To conform with the introduction of SI units, the units of activity, s^{-1}, has been given the special name of the becquerel (Bq) where $1\ \mathrm{Bq} = 1\ \mathrm{s}^{-1}$. Useful relationships between the becquerel and 'historic units' are

$$1\ \mathrm{Ci} = 3.7 \times 10^{10}\ \mathrm{Bq} \qquad 1\ \mathrm{Bq} = 2.703 \times 10^{-12}\ \mathrm{Ci}$$

$$1\ \mathrm{mCi} = 37\ \mathrm{MBq} \qquad 1\ \mathrm{MBq} = 0.0270\ \mathrm{mCi}.$$

Radioactive decay occurs spontaneously, thus the activity of any nuclide varies exponentially with time by the relationship

$$A = A_0 \exp(-\lambda t) \tag{1.1}$$

where A_0 is the initial activity and λ is the decay or transformation constant. This constant is defined for any nuclide to be the quotient of $\mathrm{d}P$ by $\mathrm{d}t$ where $\mathrm{d}P$ is the probability of a given nucleus undergoing a spontaneous nuclear transition from that energy state in the time interval $\mathrm{d}t$, that is

$$\lambda = \mathrm{d}P/\mathrm{d}t.$$

In practice, λ is related to the half-life of the nuclide, $T_{1/2}$, which is the time taken for the activity A to reach one half of the initial value A_0. By substituting $A = (1/2)A_0$ in equation (1.1), we have

$$(1/2)A_0 = A_0 \exp(-\lambda T_{1/2})$$

therefore

$$\lambda T_{1/2} = \ln(2)$$

$$T_{1/2} = 0.693/\lambda. \tag{1.2}$$

When the half-life of a nuclide is long compared with the treatment time, the term milligram–hour, suggested by Turner (1909) for radium, could be used. To cope with the variations of activity during treatment time due to decay, as is the case with radon, the unit 'millicurie destroyed' (MCD) was suggested by Debierne and Regaud (1915) as a suitable clinical unit for radon dosage. In order to relate mg h of radium to MCD of radon when considering the dose to a patient, use is made of the concept of the average life of an isotope. This defines the period in which the number of disintegrations, if they continue at the initial rate, would equal the number of disintegrations which would occur in an infinite period of time, figure 1.1.

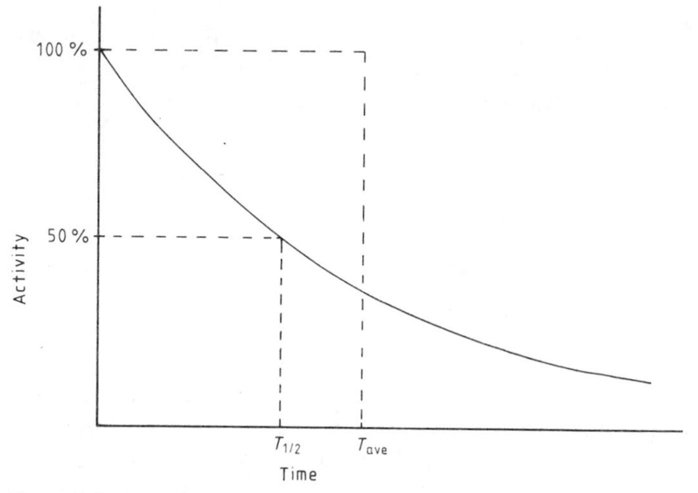

Figure 1.1 Variation of activity with time showing half-life, $T_{1/2}$ and average life, T_{ave}.

By definition the initial activity equals $N_0\lambda\ s^{-1}$, where N_0 is the initial number of atoms and λ is the probability that one atom will decay per second. In an infinite period of time all N_0 atoms will decay and this will be equivalent to the number of atoms which decay during the average life of the nuclide, T_{ave}. Thus

$$N_0 = N_0\lambda T_{ave}$$

therefore

$$T_{ave} = 1/\lambda.$$

Using equation (1.2),

$$T_{ave} = T_{1/2}/0.693$$

$$T_{ave} = 1.443T_{1/2}. \qquad (1.3)$$

This means that if any radionuclide could continue to decay at a constant rate equal to its initial rate for a time $1.443T_{1/2}$, then it would undergo the same number of disintegrations as if it decayed exponentially to infinite time with a half-life of $T_{1/2}$.

In 1932 Sievert proposed another unit for clinical use based on activity. This was the 'intensity millicurie–hour' which was the intensity at 1 cm distance from a radium preparation containing 1 mg of the radium element (as pure radium sulphate) in equilibrium with its breakdown products when the source of radiation is surrounded in all directions by 0·5 mm of platinum and the preparation enclosed by the filter is so small that the source may be considered to be a point. This unit of dose, in terms of the amount of the nuclide present, became established as a means of quantifying the dose given to a patient.

1.2.3 Exposure

Whilst some workers attempted to measure radiation in terms of its activity, others began to quantify it in terms of its ability to ionise air, exploiting the fact that a single ionising particle releases a substantial electric charge. Since it was convenient to collect ions released in a gas, air-filled ionisation chambers began to be used to measure radiation. In 1908 Villard defined a unit, the Villard unit, to be 'that quantity of radiation which liberates ionisation of one electrostatic unit (ESU) of electricity per cm^3 of air under normal conditions of temperature and pressure'. Failla in 1917 suggested defining a gamma-ray dose unit by means of the ionisation at 1 cm from a 1 g radium source.

Many other units, such as the D unit of Mallett and the Solomon R unit, also evolved during this period but these were ultimately changed following the definition of the roentgen by the Second International Congress of Radiology in Stockholm in 1928. This unit was initially defined for x-radiation but in 1931 Mayneord suggested that gamma-rays could also be measured in the same units. This concept was incorporated into the redefinition of the roentgen in 1937. However, the problems of the quantity of radiation being measured led to confusion. This was overcome when the definition was modified by the 1956 meeting of the ICRU (see ICRU (1957)) to define the roentgen in terms of the exposure dose of x- or gamma-radiation at a certain place as being a measure of radiation that is based upon its ability to produce ionisation. To avoid confusion the word 'dose' was omitted from the definition of the roentgen in 1962 by the ICRU. Various modifications were again made to the definition of the roentgen by the ICRU in 1968 and 1971 (ICRU 1968, 1971) and the latest definition of exposure (ICRU 1980) states that

> The exposure X is the quotient of dQ by dm where the volume of dQ is the absolute value of the total charge of the ions of one sign produced in air when all the electrons (negatives and positives) liberated by photons in air of mass dm are completely stopped in the air.

That is,

$$X = \mathrm{d}Q/\mathrm{d}m \quad \mathrm{C\,kg^{-1}}.$$

The special unit of exposure, the roentgen (R), which may be used temporarily is defined to be

$$1\ \mathrm{R} = 2.58 \times 10^{-4}\ \mathrm{C\,kg^{-1}}\ \text{(exactly)}.$$

1.2.4 Absorbed dose

Whereas activity related the amount of radiation a patient received to the time for which a given quantity of nuclide was used, and exposure defined 'dose' in terms of ionisation in air, neither was a measure of the actual energy deposited in the tissue by ionising radiation. As early as 1904 Beclere recognised that the quantity of radiation absorbed in tissues was a major factor in radiology but it was not until 1950 that the ICRU felt it necessary to give consideration to correlating the dose of any ionising radiation to its biological or related effects. Following this meeting, which expressed dose in terms of the quantity of energy absorbed per unit mass (ergs per gram), the ICRU (1954) defined the term 'absorbed dose' and said

> The rad is the unit of absorbed dose where 1 rad = 100 ergs/gram.

Subsequent ICRU publications emphasised the need to consider the absorbed dose within tissue (ICRU 1957, 1961, 1962, 1971) and currently the absorbed dose D is defined by ICRU (1980) to be the quotient of dE by dm where dE is the mean energy imported by ionising radiation to matter of mass dm, i.e.

$$D = \mathrm{d}E/\mathrm{d}m.$$

The special name for the unit of absorbed dose is the gray (Gy) where 1 Gy = 1 J $\mathrm{kg^{-1}}$. (The original special unit of absorbed dose, the rad, is related to the gray such that 1 Gy = 100 rad.)

In brachytherapy the absorbed dose a patient receives is often derived from a knowledge of the exposure due to a given activity of a radionuclide, together with a factor which relates exposure to absorbed dose. Under conditions of electronic equilibrium, the energy deposited by an exposure to radiation is related to the absorbed dose in air by the factor W_{air}/e where W_{air} is the mean energy required to produce an ion pair in air. Thus, the absorbed dose in air D_{air} for an exposure X is

$$D_{air} = X(W_{air}/e). \tag{1.4}$$

Since the energy absorbed per unit mass of any medium is proportional to the mass energy absorption coefficient (μ_{en}/ρ), then the absorbed dose in the medium, D_{med}, is related to the absorbed dose in air by

$$D_{med} = D_{air}[\mu_{en}/\rho]^{med}_{air} \tag{1.5}$$

where $[\mu_{en}/\rho]^{med}_{air}$ is the ratio of the mass energy absorption coefficient for the medium to the absorption coefficient for air.

By using equation (1.4) and substituting for D_{air}, equation (1.5) becomes

$$D_{med} = X(W_{air}/e)[\mu_{en}/\rho]^{med}_{air}. \tag{1.6}$$

Now $W_{air}/e = 33.85\ \mathrm{J\,C^{-1}}$ (ICRU 1979), thus for exposure measured in the SI units of $\mathrm{C\,kg^{-1}}$

$$D_{med} = (33.85[\mu_{en}/\rho]^{med}_{air})X \quad \mathrm{J\,kg^{-1}} \tag{1.7}$$

or as more commonly presented where exposure is measured in roentgens

$$D_{med} = (0.873[\mu_{en}/\rho]^{med}_{air})X \quad \mathrm{cGy}. \tag{1.8}$$

The terms in the round brackets in equations (1.7) and (1.8) are factors which convert exposure to absorbed dose. The factor in equation (1.8) is usually represented by the factor F, the roentgen-to-rad factor.

More recently the absorbed dose in tissue has been derived from the concept of kerma, an acronym for kinetic energy released per unit mass. This concept, introduced by the ICRU (1962) to highlight the two-stage process in transferring energy to matter from indirectly ionising particles, is currently defined (ICRU 1980) as

> the quotient of dE_{tr} by dm where dE_{tr} is the sum of the initial kinetic energies of all charged ionizing particles liberated by uncharged ionizing particles in a material of mass dm.

That is,

$$K = \mathrm{d}E_{\mathrm{tr}}/\mathrm{d}m.$$

The special name for the unit of kerma is the gray (Gy).

Since the charged particles produced in the interaction process slow down and give up most, if not all, of their energy to the medium, then under conditions of electronic equilibrium the absorbed dose is related to the kerma by the relationship

$$D = K(1 - g) \tag{1.9}$$

where g is the fraction of secondary electron energy lost in bremsstrahlung production.

Air kerma is now to be used instead of exposure in radiation dosimetry, thus we can say from equation (1.9) that for a given air kerma, K_{air}, the absorbed dose in air, D_{air}, is given by

$$D_{\mathrm{air}} = K_{\mathrm{air}}(1 - g). \tag{1.10}$$

Using equations (1.5) and (1.10) we can now determine the dose in a medium in terms of kerma, i.e.

$$D_{\mathrm{med}} = K_{\mathrm{air}}(1 - g)[\mu_{\mathrm{en}}/\rho]^{\mathrm{med}}_{\mathrm{air}}.$$

Since

$$\mu_{\mathrm{en}}/\rho = (\mu_{\mathrm{tr}}/\rho)(1 - g)$$

where μ_{tr}/ρ is the mass energy transfer coefficient, then

$$D_{\mathrm{med}} = K_{\mathrm{air}}(\mu_{\mathrm{en}}/\rho)/(\mu_{\mathrm{tr}}/\rho).$$

For sources used in brachytherapy g is less than 0.5%, thus

$$(\mu_{\mathrm{en}}/\rho)_{\mathrm{air}} \simeq (\mu_{\mathrm{tr}}/\rho)_{\mathrm{air}}$$

therefore

$$D_{\mathrm{med}} = K_{\mathrm{air}}[\mu_{\mathrm{en}}/\rho]^{\mathrm{med}}_{\mathrm{air}}. \tag{1.11}$$

A comparison of equations (1.7), (1.8) and (1.11) shows that the dose in a medium is the product of $[\mu_{\mathrm{en}}/\rho]^{\mathrm{med}}_{\mathrm{air}}$ and a constant factor which is dependent on either air kerma measured in gray or exposure measured in C kg^{-1} or R. Using these relationships, conversion factors, table 1.1, have been derived which relate the absorbed dose in water, bone and muscle to exposure and air kerma for the various nuclides used in brachytherapy. It will be noted that for photons of energy between

200 keV and 2 MeV the term in the square brackets in equation (1.11) is numerically equal to 1.11 for water. Thus, with the exception of iodine-125, the absorbed dose in water, D_w, is given by

$$D_w = 1.11 K_{air}.$$

Table 1.1 Factors relating absorbed dose to exposure and air kerma. (From Wyckoff (1983).)

Nuclide	Mean photon energy (MeV)	Water A†	Water B‡	Water C§	Bone A	Bone B	Bone C	Muscle A	Muscle B	Muscle C
$^{60}_{27}Co$	1.25	37.6	0.971	1.112	35.9	0.927	1.062	37.3	0.962	1.102
$^{125}_{53}I$	0.028	34.3	0.884	1.013	164.0	4.24	4.850	35.6	0.918	1.051
$^{137}_{55}Cs$	0.662	37.6	0.971	1.112	36.0	0.928	1.063	37.3	0.962	1.102
$^{192}_{77}Ir$	0.370	37.6	0.971	1.112	36.3	0.935	1.072	37.3	0.962	1.102
$^{198}_{79}Au$	0.420	37.6	0.971	1.112	36.2	0.932	1.068	37.3	0.962	1.102
$^{226}_{88}Ra$	0.780	37.6	0.971	1.112	35.9	0.927	1.062	37.3	0.962	1.102

† Factor A converts exposure measured in $C\,kg^{-1}$ to absorbed dose in Gy.
‡ Factor B converts exposure measured in R to absorbed dose in cGy; the F factor.
§ Factor C = $[\mu_{en}/\rho]^{med}_{air}$; it relates absorbed dose to air kerma.

For air we can see from equation (1.10) that the absorbed dose in air is approximately equal to the air kerma since g is small. Since an absorbed dose of 1 Gy in air is equivalent in 100/0.873 = 114.5 R, an air kerma of 1 Gy is equivalent to an exposure of 114.5 R.

A fuller review of the relationship between exposure, air kerma and absorbed dose has been given by Greening (1981, 1985) and Dutreix *et al* (1982).

1.3 Ionising Radiation and Cells

1.3.1 Biological effects of radiation

When photons interact with matter, secondary electrons are produced by one or other of the three interaction processes, the photoelectric effect, the Compton effect or by pair production. The secondary electrons produced, in slowing down, give up their energy to the medium, thus causing energy to be absorbed by the tissue. The effect, therefore, of ionising radiation on a population of cells can be categorised in one of three ways, as follows.

(i) The ionising event may not have occurred in a critical site, thus the cells are undamaged.
(ii) The ionisation may kill the cell by depositing energy within the critical site in such a way as to prevent it from reforming. This is termed lethal damage.
(iii) The ionisation has only partially taken place within the critical site, thus causing damage to the cell without killing it. This is termed sub-lethal damage since with time the cell may be able to repair this damage and completely recover from the effects of the radiation.

The magnitudes of these effects are a function of the dose. If the surviving fraction of cells is plotted against dose, on a semilogarithmic plot, then one observes a broad shoulder followed by a steeper and straighter portion, figure 1.2. The shape of this curve is often defined by two parameters.

(i) The first parameter is the slope of the final straight portion of the curve from which can be deduced the dose, D_0, which is required to reduce the number of clonogenic cells to 37%, that is $1/e$, of their former value; a clonogenic cell being one which survives a given dose of radiation and still retains its reproductive integrity and ability to proliferate indefinitely to produce a large clone or colony of cells.
(ii) The second parameter is the extrapolation number, n, which is found by extrapolating the straight portion of the curve until it intersects the 'surviving' fraction axis. This number provides a measure of the extent of the initial shoulder of the curve.

Another parameter often quoted and derived from the curve is the quasi-threshold dose, D_q, such that

$$\ln n = D_q/D_0.$$

If instead of giving the dose in one fraction it is divided into a number of equal fractions, it is found that the shoulder of the survival curve is repeated following each fraction, provided the time interval between successive fractions is sufficient for repair to take place. Further study of this type of fractionated cell survival curve shows that if single points,

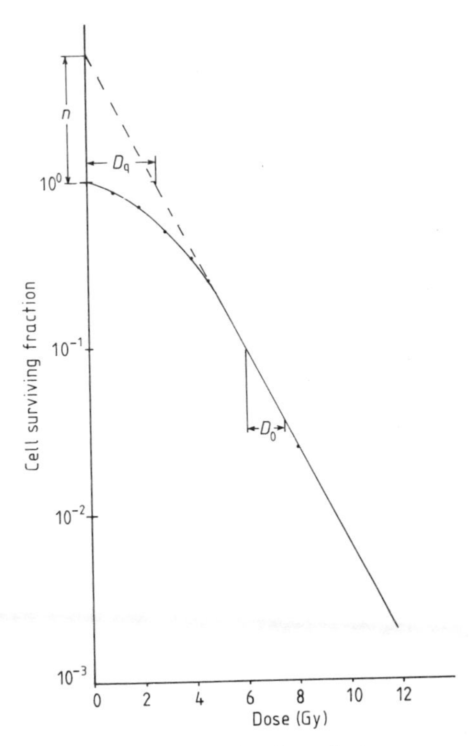

Figure 1.2 Typical cell survival curve.

corresponding to equal dose increments, are plotted on a semi-logarithmic scale, then the overall survival curve is a straight line with no shoulder. Since continuous low-dose-rate irradiation may be considered to be an infinite number of small fractions, it follows that the survival curve for this situation would therefore be expected to be an exponential function. Using this reasoning, Lathja and Oliver (1961) predicted that as the dose rate is reduced the survival curve becomes progressively less steep with the extrapolation number tending to unity and that as the dose rate is lowered and the treatment time protracted, more and more sub-lethal damage is repaired during the exposure until a dose rate is reached where all sub-lethal damage is repaired. At this point there can be no further dose-rate effect. These predictions were first substantiated by Berry and Cohen (1962) and subsequently by Hall and Bedford (1964) using mammalian (Hela) cells, cultured *in vivo*, and by Fox and Nias (1970) using Chinese hamster (ovary) cells. The way in which the slope and extrapolation number vary with changing dose rates is illustrated in figure 1.3. This shows that the dose-rate effect is clearly most important between 0.60 Gy h^{-1} and 60 Gy h^{-1}.

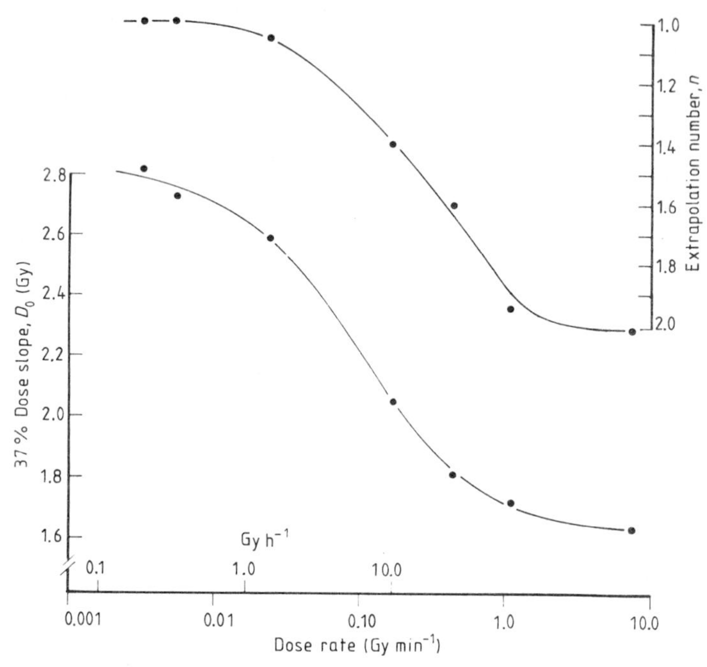

Figure 1.3 Variation in 37% dose slope, D_0, and extrapolation number, n, with dose rate. (From Hall (1972).)

It has also been established that for continuous irradiation, over a number of cell generations, at a dose rate of the order of 0.2 Gy h^{-1}, the cell cycle time of a given population becomes an important parameter in cell survival. For example, a dose rate of 0.19 Gy h^{-1} produces an effect on Hela cells, with a cycle time of 24 hours, equivalent to a dose rate of about 0.34 Gy h^{-1} in Chinese hamster cells whose cell cycle time is 10 hours. If the dose rate is further reduced to 0.10 Gy h^{-1}, then a steady state situation occurs where the death rate per cell cycle does not exceed birth rate (Hall 1972). Since normal tissues tend to have a long cell cycle time compared to tumours and much longer than Hela cells *in vivo*, one can draw the conclusion that the usual dose rates used in brachytherapy, at least in the immediate vicinity of the sealed sources, are sufficient to inhibit cell division and that for a given dose tumour cells are subject to more killing than normal tissue, thereby implying a favourable therapeutic gain.

Evidence also exists to support the theory that protracted radiation is more effective because the oxygen enhancement ratio (OER) is reduced at low dose rates, the rationale for this being that sub-lethal damage proceeds faster in aerobic than hypoxic cells. This effect was first demonstrated by Hall *et al* (1966a) when they irradiated Hela cells with cobalt-60 radiation at a continuous dose rate of 0.32 Gy h^{-1}. The study showed that the dose survival curves under oxygenated and extreme hypoxic conditions approximated to single exponential curves with values of D_0 of 2.6 Gy and 4.1 Gy respectively. This gave an OER of approximately 1.5 compared with 2.4 for acute irradiation of these cells at 1.1 Gy m^{-1}. Further evidence supporting this theory was published by Fox and Nias (1970) who measured an OER of 3.2 with acute radiation, whereas for continuous irradiation with dose rates of the order of 1 Gy h^{-1} the OER was only 1.4. Techniques for measuring OERs raise many technical difficulties which resulted in variations in reported values of OERs. Values recently reviewed by Hall and Rossi (1975) indicate that the OERs for radium-226 or caesium-137 are in the range 2.0–2.5 for dose rates of 0.44 Gy h^{-1} to 0.6 Gy h^{-1}.

The biological basis for using low-dose-rate irradiation from sources having long half-lives has been postulated by Cohen (1950, 1968), particularly in relation to permanent implants of iodine-125. The basis for the therapeutic advantage of long half-life with permanent implants lies in the differential recovery rates between tumour and normal tissue derived from the iso-effect data on skin tolerance dose and tumour lethal dose of skin cancers. From these iso-effect data, Cohen postulated that iodine-125, for example, produces a dose–time relationship which lies in the most effective region between the tissue tolerance and the minimum tumour lethal dose. Since there is a high therapeutic ratio, it may be concluded that the precision required for dosimetry is less critical than for radio-isotopes with shorter half-lives.

Brachytherapy by low-dose-rate, continuous irradiation has been considered by many radiotherapists to be the treatment of choice 'par excellence' for a number of clinical situations (Andrews 1968). The rationale for such an opinion is based on the facts that are detailed below.

(i) Due to steep dose gradients around implanted sources, normal tissues are irradiated at a lower dose rate and receive less total dose than the tumour. This causes greater cell killing in tumours, which are growing more rapidly in comparison with the slower cell cycle of normal tissue.
(ii) The dose rates used are in the range where sub-lethal damage is repaired during the exposure.
(iii) The low OER makes brachytherapy advantageous.

In recent years a neutron sealed source, californium-252, has been available for brachytherapy. In 1965 Schlea and Stoddard suggested the use of this radionuclide as a replacement for radium, the reasoning for this being based on the expectation that the OER for the mixed neutron and gamma-ray emission from californium-252 would be better than that for radium gamma-rays (Schlea and Stoddard 1965). Neutrons interact with atomic nuclei rather than with electrons and these interactions are, for the most part, elastic scattering collisions in which any kinetic energy lost by the neutron appears as kinetic energy in the recoil of the nucleus which has a higher linear energy transfer. The overall effect, therefore, is that cell killing for fast neutrons is less oxygen dependent than for gamma-rays, although other effects such as dose-rate differences complicate the direct comparisons of gamma-ray therapy using radium or its substitutes and the mixed neutron and gamma-ray therapy from californium-252. Measurements have shown that the OER for the mixed neutron and gamma-rays from californium-252 at low dose rates (in the range 0.15–0.20 $Gy\,h^{-1}$) is in the range 1.3–1.7 compared with the OER for gamma-rays which is in the range 2.0–3.0. The reported values of the relative biological effectiveness (RBE) of the mixed radiation dose from californium-252 compared to radium-226 have been also summarised by Hall and Rossi (1975) and are typically in the range of 3.5–8, the value being a function of dose rate. By incorporating the OER considerations and taking a value of 7 for the RBE for the mixed neutron and photon beam from californium-252, Hall (1972) has suggested that for clinical practice '8.9 Gy of "Californium neutrons", given in 7 days would be expected to give the same response as 60 Gy of radium gamma rays in seven days.' This corresponds to a neutron dose rate of 0.053 $Gy\,h^{-1}$, which with the gamma-rays gives an overall dose rate of 0.10 $Gy\,h^{-1}$.

1.3.2 Brachytherapy dose–time relationships

Decades before radiobiologists were able to study the effect of low-dose-rate, continuous irradiation on cells, radiotherapists were gaining experience in the use of sealed sources in brachytherapy. Regaud, working at the Foundation Curie, Paris, in the 1920s showed that prolonged low-dose-rate irradiation was beneficial. The principles he founded of using relatively small quantities of radium implanted for periods of from six to ten days are still followed, albeit in many instances with radium substitutes. Early workers also began, clinically, to find the relationship between dose and time. One of the most frequently used dose–time relationships used for brachytherapy stemmed from the work of Paterson and his colleagues at the Christie Hospital and Holt Radium Institute, Manchester, in the 1930s. Experience showed that, depending on the area/volume implanted, doses in the range 5500–7000 R in six to eight days could be given which would produce tumour control without exceeding the tolerance of normal tissues. (Initially the unit of dose was based on the Paterson–Parker 'roentgen', which was later shown to be equivalent to the roentgen.) Whilst this represented the tolerance dose, experience also showed that the maximum dose which could be given varied with dose rate. If the dose rate was higher or lower than pre-calculated, due to the actual volume implanted being smaller or larger than estimated, then it became apparent that the dose given should be either reduced or augmented. This led Paterson (1963) to publish his iso-effect corrections which stated that

> for every 24 hours which the time calculated from the actual dose rate falls short of, or exceeds the intended 7 day interval, the time to be prescribed must be reduced or increased respectively by eight hours.

The corrected time to give doses equivalent to 7000 R in 168 hours can therefore be expressed by the formula

$$\text{corrected time} = H - (168 - H)/3 \quad \text{hours}$$

where $H = 7000/\text{dose rate}$ within the implant in R h^{-1}, or for an equivalent of 60 Gy in seven days, Wilkinson (1972) described the formula

$$\text{corrected time} = (8000/d) - 56 \quad \text{hours}$$

where d is the dose rate in cGy h^{-1}.

This simple rule was shown through clinical experience to be a good approximation for treatment times in excess of four days. Similar data due to Green (unpublished) and cited by Ellis (1963, 1968) showed a similar dose–time relationship (figure 1.4). This dose–time relationship has been represented by Ellis to be of the form

$$D_1 = D_2(T_1/T_2)^{0.26}$$

where D_i is the dose given in time T_i, $i = 1, 2\ldots$, and by Wilkinson (1972) to be

$$D_r = D_7(d_1/d_2)^{0.45}$$

where D_r is the dose required, D_7 is the equivalent seven-day dose, d_1 is the dose rate required to give D_7 and d_2 is the dose rate achieved in the particular case.

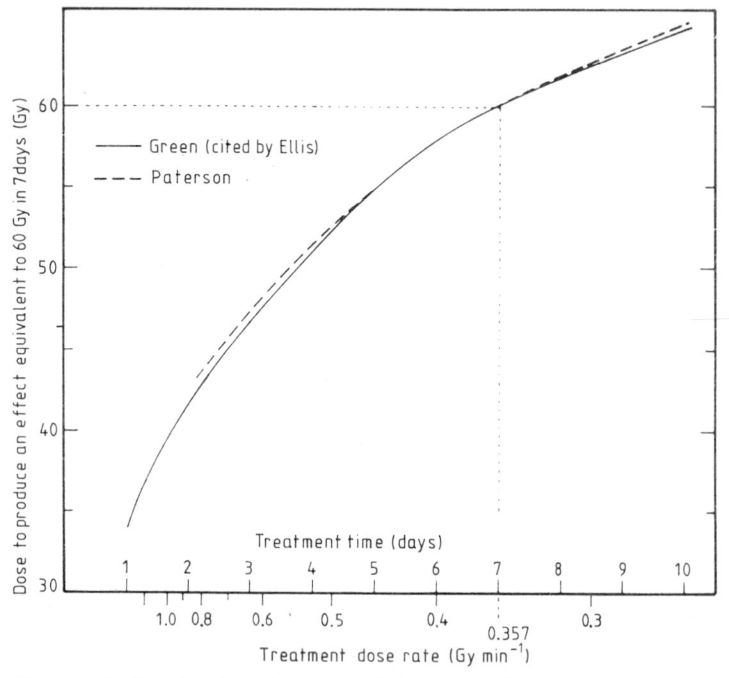

Figure 1.4 Relationship of the total dose to overall treatment time for an implant to produce an effect equivalent to 60 Gy in seven days. (From Hall (1972).)

Orton (1974) has also used these data, together with other sources, to extend the time–dose factor concept (TDF), developed for fractionated radiotherapy, to brachytherapy. This concept provides a method for comparing and, if necessary, modifying brachytherapy regimes to account for different dose rates. By plotting the iso-effect curve on a log–log scale and using least square regression analysis, it was found that treatment times for different dose rates could be represented by the expression

$$T = (2.10 \times 10^4)r^{-1.35} \tag{1.12}$$

where r is in cGy h^{-1} and T in hours, or

$$T = 41.9r^{-1.35} \tag{1.13}$$

where r is in Gy h^{-1}.

For brachytherapy the definition of TDF is

$$\text{TDF} = K(T/T_{\text{tol}}) \tag{1.14}$$

where K is the TDF for full tolerance, T is the actual treatment time at that dose rate, and T_{tol} is the time of application at a chosen dose rate which would result in full tolerance. By equating TDF to 100 for a dose of 60 Gy in seven days and substituting the value for T from equation (1.12) in equation (1.14) we obtain the expression

$$K/(T_{\text{tol}})_r = 100/[(2.10 \times 10^4)r^{-1.35}]. \tag{1.15}$$

Substituting this expression in the TDF equation (1.14) we have

$$\text{TDF} = (4.76 \times 10^{-3})r^{1.35}\,T \qquad \text{where } r \text{ is in cGy h}^{-1} \tag{1.16}$$

$$\text{TDF} = 2.39r^{1.35}\,T \qquad \text{where } r \text{ is in Gy h}^{-1}. \tag{1.17}$$

This is the expression for a constant dose rate, r. However, r is often a function of time and since TDFs are additive and linearly proportional to time, the final TDF reached in time T is the integral sum of TDFs for each interval of time. Thus the general equation for TDF is

$$\text{TDF} = 2.39{r_0}^{1.35}\,T_{\text{eq}} \tag{1.18}$$

where r_0 is the initial dose rate and T_{eq} is given by

$$T_{\text{eq}} = [1 - \exp(-1.35\lambda T)]/1.35\lambda. \tag{1.19}$$

For long-lived isotopes where $\lambda \rightarrow 0$, $T_{\text{eq}} \rightarrow T$ and thus the TDF equation reduces to equation (1.17). For the temporary application of short-lived isotopes, the appropriate value of T_{eq} must be used, whilst for the permanent application of short-lived isotopes where $T \rightarrow \infty$, equation (1.19) reduces to

$$T_{\text{eq}} = 1/1.35\lambda. \tag{1.20}$$

Using these equations Orton has devised TDF tables for both the temporary and permanent application of isotopes in brachytherapy (Orton 1974). These tables substantiate the accepted rule that because the biological effect of radioactive gold is greater than radon due to its shorter half-life, the dose delivered in a permanent implant per unit activity of gold will be approximately 10% greater than per unit activity of radon. For example, to achieve the equivalent of 60 Gy in seven days (TDF = 100) for a permanent radon-222 implant, an initial dose rate of 0.53 $\mathrm{Gy\,h^{-1}}$ is required. This gives a total dose of $0.53 \times 132.6 = 70.3$ Gy. If gold-198 seeds are used, the tables indicate an initial dose rate of 0.69 $\mathrm{Gy\,h^{-1}}$ to achieve the same biological effect. In this case the total dose is $0.69 \times 93.2 = 64.3$ Gy.

Another approach to assessing the biological effects of fractionated regimes and brachytherapy was developed by Kirk *et al* (1971, 1972, 1973, 1975a,b, 1977a,b). These assessments are based on a scale of accumulated sub-tolerance radiation damage, the cumulative radiation effect (CRE). This work is comparable with that of Orton, although account is also taken of the volume irradiated. For continuous irradiation the CRE, in radiation effective units (REU), is given by

$$R_c = \mu q \phi r T^{0.71}$$

where $\mu = 0.77$, q is the RBE factor (for radium RBE factor = 1.02, for caesium RBE factor = 1.03 and for iridium RBE factor = 1.06 (Kirk *et al* 1972)), ϕ is the volume factor such that $\phi = (V/1000)^{0.16}$ (V is the volume), r is the dose rate in cGy per day and T is the time in days. A comparison between the TDF and CRE systems has been made by Goitein (1976) and Orton (1980) who has recommended that the value of μ should be 0.53 as originally suggested by Kirk *et al* (1972).

In recent years *in vivo* cell survival curves have increasingly been described by the linear quadratic cell survival equation

$$S = \exp[-(\alpha D + \beta D^2)]$$

where S is the surviving fraction of cells, D is the dose per exposure and α and β are constants related to the two processes which may lead to cell death. These processes occur

(i) when two critical targets in a cell are simultaneously hit in a single event—type A damage, represented by αD—and

(ii) when two targets are damaged by two separate radiation events which produce cell death through mutual interaction—type B damage, represented by βD^2.

If only one of the two targets is hit, then (ii) may be considered to be sub-lethal damage.

Dale (1985) has extended this linear quadratic model of dose response to protracted irradiation brachytherapy and derived a generalised equation for the relative effectiveness per unit dose. This equation can be used as a method for determining the biological equivalent dose for different brachytherapy techniques.

The clinical data of Paterson and Green, however, is not substantiated by all workers. Pierquin *et al* (1973) and Inoue *et al* (1978) were both unable to find evidence of a dose-rate effect in their interstitial techniques, indeed both groups indicated that for a dose of 70 Gy given at a dose rate of 0.25 to 1 $Gy\,h^{-1}$, i.e. between three and eleven days, there was no adjustment necessary in treating oral carcinoma with iridium-192 implants.

Dose rates used in intracavitary treatments have traditionally been in excess of 0.5 $Gy\,h^{-1}$. Numerous techniques have evolved over the years for the treatment of carcinoma of the uterine cervix but most are modifications of either the Stockholm technique (Heyman 1929) which originated at the Radiumhemmet in Stockholm or the Paris technique (Regaud 1929) developed at the Radium Institute in Paris. With the former the treatment is fractionated with two or three applications of about 20–24 hours per insertion over a period of three to four weeks. Dose rates to the reference point A are of the order of 1.0–1.2 $Gy\,h^{-1}$. However, the Paris system and the Manchester technique from which it evolved have dose rates of the order of 0.54 $Gy\,h^{-1}$ at the reference point. In this approach the fractionation consists of two insertions of three days, separated by a four-day gap.

In recent years after-loading techniques have been used which operate at much higher dose rates and this has required work to be carried out to establish the equivalent dose–time relationships. These techniques fall into two main categories, dose rates of the order of 1.75 to 2 $Gy\,h^{-1}$ and higher rates of 4 $Gy\,min^{-1}$. Dose–time relationships for the former technique have been published by Wilkinson *et al* (1980) using a Selectron after-loading system, whilst the use of the higher dose rates afforded by the Cathetron have been studied by numerous authors who proposed a general formula for equating protected and acute regimes of radiation (O'Connell *et al* 1967, Joslin and Smith 1970, Liversage 1969). If a low-dose-rate technique employing continuous irradiation for t hours is replaced by a high-dose-rate fractionated regime of N fractions, then the number of fractions which will result in the same biological response will be given by

$$N = \mu t/2[1 - (1 - e^{-\mu t})/\mu t]$$

where $\mu = 0.693/T$ and T is the half-life of the fading of the dose equivalent of sub-lethal dosage which is typically in the range 1 to 1.5 hours. Using this value of T, for $t > 12$ hours this expression approximates to $N = t/4$. This approach has been the basis of much of the experience gained by the use of high dose rates in intracavitary therapy where treatment, to be equivalent to 40 Gy in 48 hours, might consist of six fractions of 5.5 Gy to the reference point in an overall time of twelve days.

2 Production and Construction of Sealed Sources

2.1 Brachytherapy Source Requirements

For many years brachytherapy was performed using radium and its daughter element radon. Whilst the former has the advantage of a very long half-life, it has the disadvantage of producing an alpha-emitting gaseous daughter product which is soluble in tissue and highly radiotoxic. Should any of the isotope leak from the sealed container due to mechanical damage or accidental incineration, a major hazard arises. In addition, the gamma-radiation from these isotopes is of high energy and thus large thicknesses of lead or other protective materials are needed for safe handling. These disadvantages led to the introduction in the late 1940s of substitutes for radium brachytherapy. The choice of isotopes which can be used, however, is limited since few have all the desirable properties of the 'ideal' brachytherapy source. These properties are as follows.

(i) The optimum gamma-ray emission should be high enough to avoid increased energy deposition in bone by the photoelectric effect and also high enough to minimise scatter, whilst at the same time be low enough to minimise protection requirements. To satisfy all the requirements the optimum energy should be in the range 0.2 to 0.4 MeV.

(ii) The half-life should be such that correction for decay during treatment is minimal. In addition, for permanent stock a very long half-life is desirable so that radioactive decay within the lifetime of the source and its container is negligible and stock can be easily used and stored.

(iii) Charged-particle emission should be absent or easily screened.

(iv) There should be no gaseous disintegration product.

(v) The nuclide should have a high specific activity.
(vi) The material should be available in insoluble and non-toxic form.
(vii) The material should not powder or be otherwise dispersed if the source is damaged or incinerated.
(viii) The isotope should be able to be made in different shapes and sizes including rigid tubes and needles, small spheres and flexible wires. In the case of the latter they should be able to be cut into the lengths required without danger or contamination.
(ix) Damage during sterilisation should not be possible.

2.2 Production of Radionuclides

Over the years many isotopes have been used for brachytherapy but none has satisfied all of the above criteria for the ideal source. The supply of the radionuclide comes from two main sources; those occurring in nature and those produced artificially in a nuclear reactor. Isotopes produced by the latter method occur as either a fission product within a spent uranium fuel rod or as the result of an $n-\gamma$ reaction taking place in a stable isotope which is irradiated in the neutron flux occurring within a nuclear reactor. These neutron activation reactions are of the form

$${}^{A}_{Z}X + {}^{1}_{0}n = {}^{A+1}_{Z}X + \gamma$$

The gamma-ray produced at the time of irradiation has an energy due to differences in mass units between parent and daughter elements; it does not have the same energy as any gamma-ray associated with the decay of the daughter element. The activity of the isotope produced in this process is dependent upon the number of parent nuclei being bombarded, the probability of a nucleus capturing a neutron (the cross section for the nuclear reaction), the neutron flux, the transformation constant of the isotope produced and the length of time the material stays in the neutron flux. These parameters not only affect the final activity of the isotope produced but also determine the activity per unit mass, the *specific activity*. It can be shown that the activity of an isotope after being bombarded in a neutron flux density, θ, for a time, t, is given by

$$A = N_t \phi \theta [1 - \exp(-0.693t/T_{1/2})] \quad (2.1)$$

where N_t is the number of nuclei being bombarded, ϕ is the probability of activation, i.e. the cross section for the nuclear reaction in cm^2 per atom or 'barns' per atom and θ is the neutron flux density expressed by neutrons $cm^{-2} s^{-1}$.

This equation indicates that the actual growth in activity is not linear with time but reaches a maximum, the *saturation activity*, after several half-lives. The product $N_t\phi\theta$ is equivalent to the saturation activity of the isotope and occurs when the rate of production of the active atoms equals the rate at which they decay. A more detailed account of this method of isotope production has been given by Johns and Cunningham (1983). A practical example of the activities produced is given below.

Example 2.1
If 10 g of iridium-191 is placed in a flux of 10^{13} neutrons $cm^{-2} s^{-1}$, what activity of iridium-192 will be produced in one year?

Since the atomic weight of iridium is 191, the number of atoms in the target N_t is given by

$$\begin{aligned} N_t &= (\text{Avogadro's number/atomic weight}) \times \text{mass} \\ &= (6.022 \times 10^{23}/191)\ g^{-1} \times 10\ g \\ &= 3.153 \times 10^{22}. \end{aligned}$$

The activation cross section for iridium-191 is 910 barns, thus the activity produced in the 10 g in one year, from equation (2.1), will be

$$\begin{aligned} A &= (3.153 \times 10^{22})(910 \times 10^{-24}) \times 10^{13}\ [1 - \exp(-0.693 \times 365 \times 1/74.0)] \\ &= 2.775 \times 10^{14}\ \text{Bq (7500 Ci)}. \end{aligned}$$

Since the initial mass of iridium-191 is 10 g, the specific activity of the iridium-192 produced is 2.775×10^{10} Bq kg^{-1} (750 Ci g^{-1}).

Table 2.1 Mass of radionuclide for given activity.

Nuclide	Mass in μg for activity of 100 MBq	Mass in mg for activity of 1 Ci
Cobalt-60	2.39	0.884
Caesium-137	31.3	11.56
Iridium-192	0.294	0.109
Gold-198	0.011	0.0041
Radium-226	2732.5	1011.0

Table 2.2 Physical properties of radionuclides used for brachytherapy and their mode of production.

Nuclide	Half-life	Type of disintegration	Maximum beta-ray energy (MeV)	Photon energy Mean (MeV)	Photon energy Maximum (MeV)	Thermal cross section for neutrons¶ (barns)	Method of production
$^{60}_{27}Co$	5.27 years	Beta, gamma	1.49	1.25	1.33	36.9	n–γ
$^{125}_{53}I$	60.1 days	Electron capture with x-ray emission, gamma	—	0.028	0.035	100	Decay product†
$^{137}_{55}Cs$	30.0 years	Beta, gamma	1.17	0.662	0.662	—	Fission product
$^{182}_{73}Ta$	115 days	Beta, gamma	0.59	0.70	1.29	22.0	n–γ
$^{192}_{77}Ir$	74.0 days	Beta, gamma	0.67	0.37	1.06	910	n–γ
$^{198}_{79}Au$	64.7 hours	Beta, gamma	0.96	0.42	0.68	98.8	n–γ
$^{222}_{86}Rn$‡	91.8 hours	Alpha, beta, gamma	3.27	0.78	2.45	—	Naturally occurring
$^{226}_{88}Ra$‡	1600 years	Alpha, beta, gamma	3.27	0.78	2.45	—	Naturally occurring
$^{90}_{38}Sr$§	28.7 years	Beta	0.546	—	—	—	Fission product
$^{90}_{39}Y$	64.0 hours	Beta	2.27	—	—	13.0	n–γ
$^{252}_{98}Cf$	2.64 years	Fission,‖ alpha, gamma	—		0.10	—	Fission

† Daughter element of $^{125}_{54}Xe$ decay, which itself is produced by n–γ reaction.
‡ Gamma photons are produced by daughter elements in equilibrium with radium and radon.
§ Used when in equilibrium with $^{90}_{39}Y$.
‖ Neutron energies in the range 2.1 to 2.3 MeV, modal energy 1.0 MeV. Other gamma-ray photons are produced during decay of daughter elements.

Similarly, it can be shown that the specific activity induced in cobalt-60 and gold-198 if left in the same neutron flux for one year will be 4.65×10^8 Bq kg^{-1} (12.6 Ci g^{-1}) and 3.02×10^9 Bq kg^{-1} (81.7 Ci g^{-1}) respectively.

One of the requirements for an ideal brachytherapy source is that the radionuclide has a high specific activity, thus enabling the production of small sources. Since activity is defined as $N_t\lambda$ Bq, we can show that the mass m, in grams, for an activity of 1 Bq is given by

$$m = (1/\lambda)A/N_a$$

where N_a is Avogadro's number which is equal to the number of atoms in a mass of material in grams numerically equal to its atomic weight, A.

For isotopes used in brachytherapy the mass of nuclide required for a given activity is detailed in table 2.1. The physical properties of the radionuclides used for brachytherapy, together with their mode of production, is given in table 2.2.

2.3 Brachytherapy Sources

2.3.1 Radium-226 sources

Radium-226 occurs in nature as one of the daughter products of the uranium series which starts with $^{238}_{92}U$ and ends with stable $^{206}_{82}Pb$. Radium-226 is produced by the disintegration of $^{230}_{90}Th$ and when isolated from its parent decays with a half-life of 1600 years to form the inert gas radon by the emission of alpha-particles and a weak gamma-ray of energy 0.18 MeV. Radon then decays by alpha-particle emission to form polonium-218. The elements in the decay series, figure 2.1, emit alpha- and beta-particles of various energies together with gamma-rays with a total of 72 different energies (Payne and Waggener 1974b). However, a large proportion of these are of low energy and thus are of little or no benefit in brachytherapy. The gamma-rays which make the largest contribution to the gamma-ray dose are those which emanate from bismuth-213 (radium C) and, to a lesser extent, from lead-214 (radium B). The maximum gamma-ray energy in the decay series is 2.2 MeV and thus the radiation from a radium source is very penetrating.

When a sample of radium is placed in a sealed container, all of the daughter products of the parent nucleus will be present after a short time, but it is some considerable time before radioactive equilibrium is established. Under these conditions elements in the decay series disintegrate at the same rate as they are being produced, thus all elements in the series decay with the half-life of radium. Similar reasoning is applied to the effective gamma-rays used in radon brachytherapy, except that in this case the elements decay with the half-life of radon.

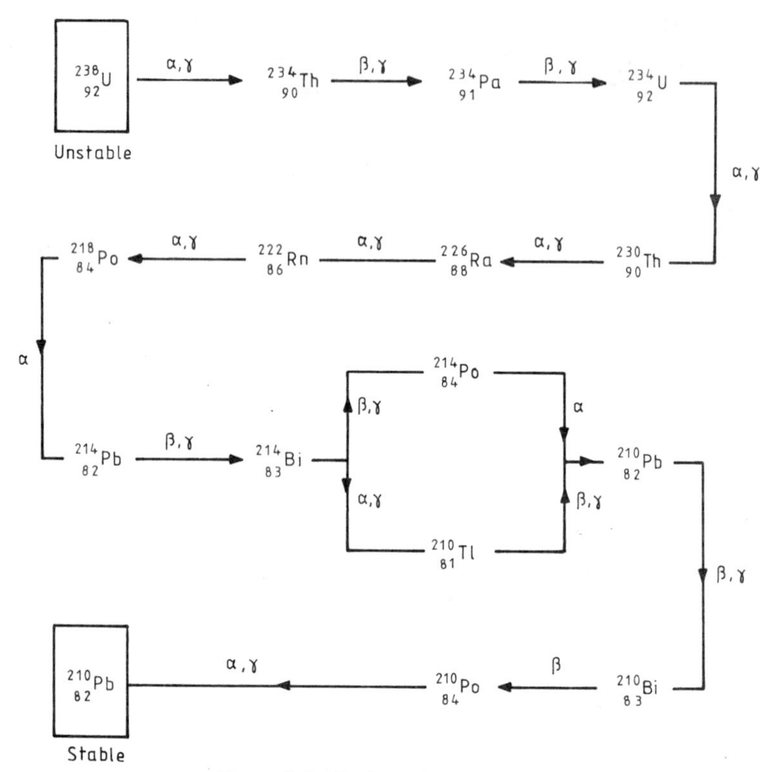

Figure 2.1 Radium decay series.

The presence of beta-particles in the decay series contributed to the necrosis experienced by early workers using radium for implant techniques, since there was insufficient wall thickness to absorb all these particles. Following the work of Neary (1942), who showed that if radium is filtered by 0.5 mm of platinum the dose contributed by the particles at the surface of the source is about 1% of that contributed by the gamma-radiation, radium sources are now constructed with sufficient wall thickness to absorb most of these energetic particles. One of the essential features, therefore, in the design of a radium source is a container made of 10% iridium–platinum alloy of wall thickness greater than 0.5 mm to absorb most of the beta-particles. In addition, in the construction of radium needles the trocar points are made of 25% iridium–platinum alloy to provide extra wear resistance.

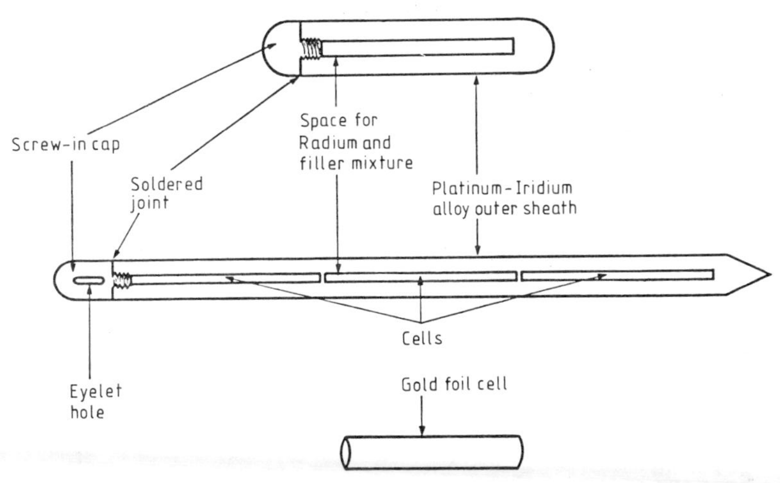

Figure 2.2 Construction of a radium-226 tube and needle with gold foil cell. (Reproduced from Meredith and Massey (1977) by kind permission of John Wright, Bristol.)

In practice, radium needles and tubes are formed by doubly encapsulating radium sulphate in platinum–iridium alloy, figure 2.2. The radioactive powder is mixed with barium sulphate and placed in a dry state within cells which are then loaded into the outer iridium–platinum container, iridium being alloyed with platinum to provide increased strength. The barium sulphate acts as a filler to ensure that the required amount of radium sulphate to produce a source of given strength is adequately dispersed within the cell, thereby preventing uneven linear activity. These cells, which are usually 1 cm long and 1 mm in diameter, are sealed by brazing and then loaded into the outer container which is itself sealed by brazing or resistance welding. This method of double encapsulation minimises the risk of a possible radium leak. The use of cells also has the advantage that should damage occur to the outer container, only a limited amount of contamination can occur. In addition, cells containing different amounts of the radium salt can be used to manufacture differentially loaded sources for implant work.

2.3.2 Caesium-137 sources

In recent years caesium-137 has begun to be used in preference to radium in brachytherapy. The source of caesium-137 is as a fission product from spent uranium fuel rods used in nuclear reactors. Difficulties in separating caesium-137 from other isotopes of caesium means that small quantities of contaminants, such as caesium-134, are often present, although this is usually less than 1% by activity. Caesium-137 decays with a half-life of 30.0 years by beta- and gamma-ray emission to the stable isotope barium-137

$$^{137}_{55}\mathrm{Cs} \rightarrow {}^{137}_{56}\mathrm{Ba} + {}^{0}_{-1}\mathrm{e} + \gamma$$

The use of caesium-137 is advantageous since it has no gaseous daughter products, its maximum photon energy is 0.662 MeV, thereby

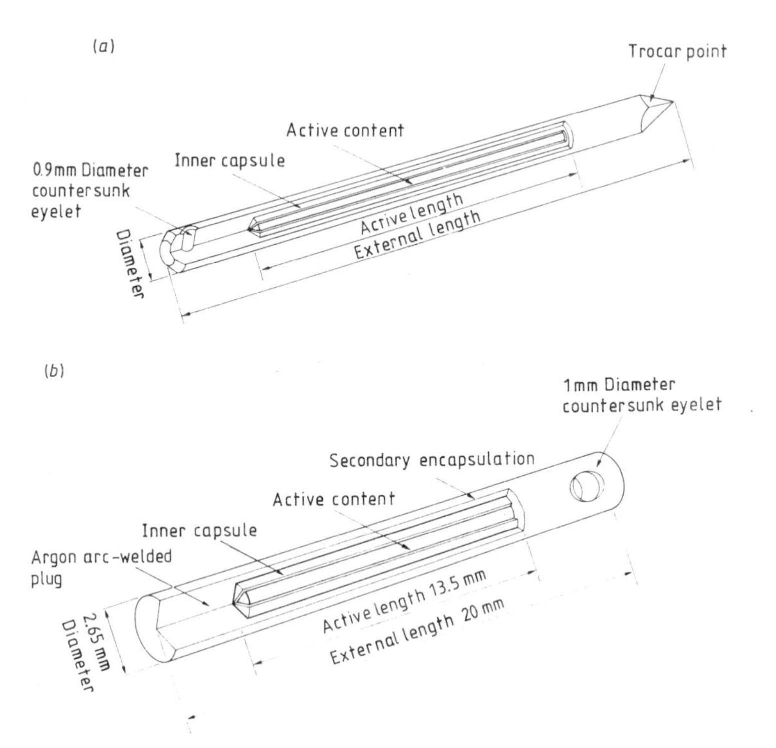

Figure 2.3 (*a*) The caesium-137 source CDCS A-type needle. (*b*) The caesium-137 source CDCS J-type tube. (Reproduced by kind permission of Amersham International plc.)

making protection easier, and its beta-rays can be reduced by a thin thickness of filtrating material. Although its half-life is much less than that of radium, the correction of 2% per year does not present major problems, but it does mean that sources need to be replaced every ten years or so.

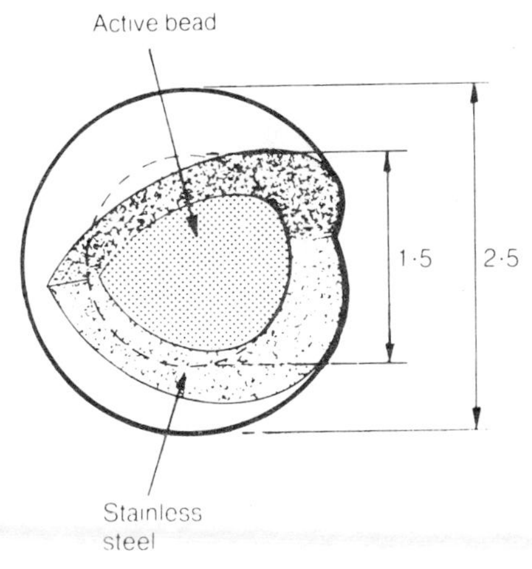

Figure 2.4 Spherical caesium-137 source as used in the Selectron after-loading system (the distances are given in mm). (Reproduced by kind permission of Nucletron Trading Ltd.)

The caesium-137 needles and tubes used as a direct replacement for radium in interstitial and intracavitary techniques are constructed using a stable insoluble compound of caesium-137 in powder form which is doubly encapsulated in iridium–platinum alloy (figures 2.3(*a*) and (*b*)). The inner cells are made of 10% iridium–platinum alloy sealed by brazing which are then placed inside an outer cover made of 10% or 20% Ir–Pt alloy sealed by resistance welding. As with the radium needles, the trocar points of needles are made of 25% Ir–Pt alloy for extra wear resistance. Since the beta-rays in caesium decay have a predominant energy of 0.514 MeV, they can be adequately filtered by a wall thickness of 0.5 mm of platinum. Recent tube and needle construction has been made using 0.5 mm of stainless steel as the outer container. The construction of after-loading sources, however, is achieved by creating caesium-137 glass beads which are then encapsulated in stainless steel by arc-welding into spherical or cylindrical forms (figures 2.4 and 2.5) which can then be made into source trains.

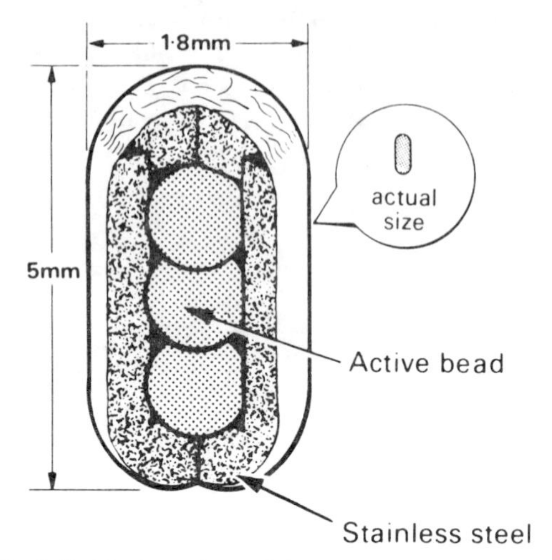

Figure 2.5 Cylindrical caesium-137 source as used in the Amersham Manual after-loading source train. (Reproduced by kind permission of Amersham International plc.)

2.3.3 Cobalt-60 sources

This radionuclide was one of the earliest artificially produced isotopes used as a radium substitute. Produced by the n–γ reaction

$$^{59}_{27}\text{Co}\ (\text{n}-\gamma)\ ^{60}_{27}\text{Co}$$

cobalt-60 decays by beta- and gamma-emission to form stable nickel-60

$$^{60}_{27}\text{Co} \rightarrow {}^{60}_{28}\text{Ni} + {}^{0}_{-1}\text{e} + \gamma$$

Since the dominant beta energy is 318 keV, it means that a thickness of 0.1 mm of platinum is sufficient to absorb these rays. This led Fletcher *et al* (1954) to fabricate thin cobalt-60 needles, by placing active wire inside a sheath of stainless steel or iridium–platinum alloy, in the hope that they would be less traumatic to tissue. However, even with the advantage of a greater exposure rate factor, high specific activity, and thinner needles, the half-life of the isotope, 5.27 years, presented problems in use since a more complex system of stock control was necessary and the sources needed replacing every two or three years. In addition, cobalt wire is brittle and difficult to sheath and this, together with the possibility of metallic cobalt being absorbed by tissue and deposited elsewhere in the body, has meant that, in practice, cobalt is a dangerous material to use interstitially because of the risk of breakage.

The use of cobalt-60 in brachytherapy tends to be restricted to ophthalmic applicators or for certain after-loading systems where the source is constructed as a 20 carat gold bead loaded with cobalt-60 wire rings around the central core and sealed with gold which acts as the beta-ray screen.

2.3.4 Iridium-192 sources

Iridium-192 has been used as a brachytherapy source since 1958, firstly in seed form by Henschke and then, since 1960, as wire and hair-pins following work at the Institute Gustave Roussy in Paris. The isotope is produced by the n–γ reaction

$$^{191}_{77}\mathrm{Ir}\ (\mathrm{n}-\gamma)\ ^{192}_{77}\mathrm{Ir}$$

However, the process also produces quantities of iridium-194 from the (n–γ) activation of iridium-193. Since this has a half-life of 17 hours compared with 74 days for iridium-192, it does not contribute any significant dose in treatment as the sources are not usually supplied until eight days after irradiation in the reactor, by which time the iridium-194 has decayed by at least ten half-lives. The unusually high cross section of iridium-191 for neutron capture (910 barns) yields a high specific activity, which means that linear activities of between 1 MBq mm^{-1} to 10 MBq mm^{-1} (30 μCi mm^{-1} to 300 μCi mm^{-1}) can be achieved. Since the cladding of the wire is slightly active due to activation of other iridium isotopes, iridium-192 wire is not a sealed source, although it is internationally accepted as a 'sealed source' because it does not contaminate cutting instruments or release active particles.

Iridium-192 decays by emitting beta- and gamma-rays to form the stable isotope platinum-192,

$$^{192}_{77}\mathrm{Ir} \rightarrow\ ^{192}_{78}\mathrm{Pt} +\ ^{0}_{-1}\mathrm{e} + \gamma$$

The beta-rays have predominant energies of 530 keV and 670 keV, whilst the main gamma-rays emitted have an average energy of 370 keV. This makes radiation protection easier.

Sources for brachytherapy are usually in the form of flexible wire of 0.3 mm diameter which can easily be cut to length as required for each application. This wire consists of an active iridium–platinum alloy core, 0.1 mm thick, encased in a sheath of platinum, 0.1 mm thick, a thickness adequate to filter out the majority of beta-rays produced in the decay schemes. These wires, which are extremely resistant and unbreakable yet highly malleable, are loaded either into plastic tubes or hypodermic needles. Iridium wire is also used in brachytherapy in the form of hair-pins or single pins (figure 2.6) which are made from wire 0.6 mm thick. Manufacturers also supply iridium-192 seeds in the form of preloaded nylon ribbons containing twelve seeds spaced at 1 cm intervals. After-loading sources can also be constructed using iridium-192—a 1 × 1 cm pellet of iridium-192 being sealed into a 2.5 mm titanium sphere for low-dose-rate techniques and into a source capsule for high-dose-rate systems.

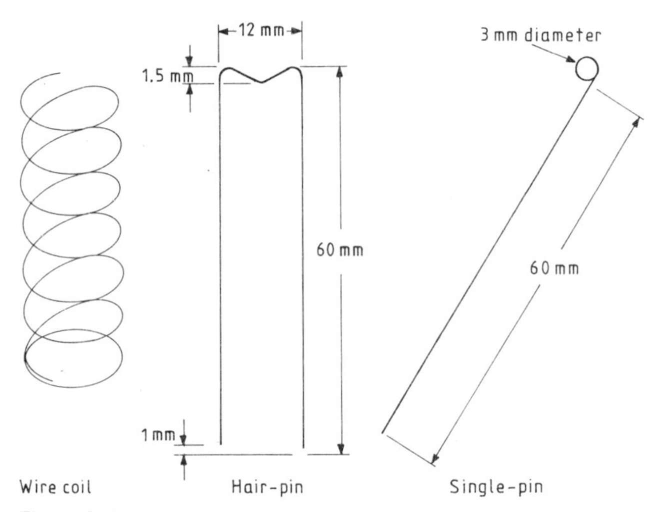

Figure 2.6 Iridium-192 brachytherapy sources. (Reproduced by kind permission of Amersham International plc.)

2.3.5 Tantalum-182 sources

The use of this radionuclide for brachytherapy was suggested as early as 1946 and in 1952 Sinclair reported the manufacture of wire and hair-pins constructed with tantalum-182. Having a half-life of 115 days and gamma-ray and maximum beta-ray energy of 1.13 MeV and 0.5 MeV respectively, this isotope was used for a number of years for interstitial techniques. In construction the form was similar to iridium wire. For clinical use wire of diameter of either 0.2 or 0.4 mm is sheathed with 0.1 mm of platinum (Sinclair 1952, Haybittle 1957). This isotope has now been replaced by iridium-192 as the isotope of choice for interstitial brachytherapy since the latter has a greater specific activity making construction easier. Protection requirements are less demanding.

2.3.6 Gold-198 sources

For permanent implantation techniques gold-198 seeds began to be used as a replacement for radon in the early 1950s (Sinclair 1952). With a high nuclear reaction cross section for neutrons, gold-198 can be produced with high specific activity by the (n$-\gamma$) reaction

$$^{197}_{79}\text{Au}\ (\text{n}-\gamma)\,^{198}_{79}\text{Au}$$

This is an important property, since for transportation purposes it is necessary to have a high initial activity because of the short half-life of 2.7 days. Gold decays by beta- and gamma-ray emission

$$^{198}_{79}\text{Au} \rightarrow {}^{198}_{80}\text{Hg} + {}_{-1}^{0}\text{e} + \gamma$$

where the maximum beta energy is 966 keV and the average gamma-ray energy is 420 keV.

Initially, gold-198 was supplied as gold wire, gold grains and gold seeds. Sources available today are usually in the form of gold seeds (figure 2.7) which are small, square-ended cylinders with a core of gold of 0.5 mm in diameter encased in 0.15 mm of platinum, a thickness sufficient to absorb the majority of the beta-particles. These grains are then loaded into magazines for use with an implantation gun (see figure 6.19). Gold has advantages over radon in that protection is easier because of the lower x-ray energy and also because the gamma-ray activity disappears after about one month, unlike radon where some of the gamma activity arising from the energetic particles of daughter

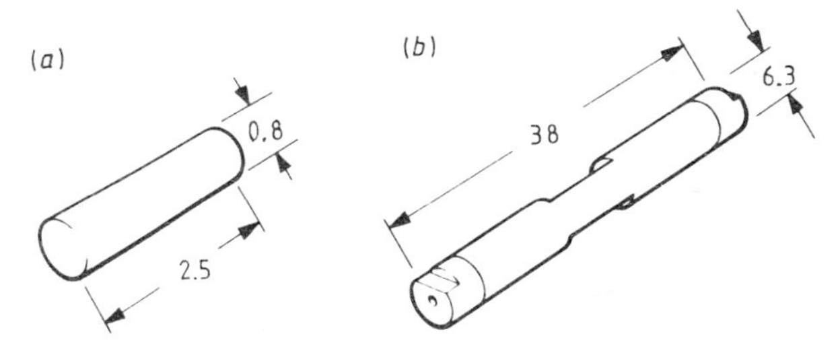

Figure 2.7 (*a*) Gold-198 grain and (*b*) magazine (distances given in mm). (Reproduced by kind permission of Amersham International plc.)

elements is still detectable after many years. The neutron activation process does, however, cause the cladding of the sources to become slightly radioactive, thus they are not classified as sealed sources.

2.3.7 Strontium-90 sources

The use of this radionuclide stems from the fact that even though it has a relatively low beta energy of 546 keV maximum, it is in radioactive equilibrium with its daughter element, yttrium-90, which has a maximum beta-ray energy of 2.27 MeV. This means that even though the half-life of yttrium-90 is only 64 hours, the more energetic beta-rays are present for the half-life of strontium-90, 28.7 years. The decay processes involved are

$$^{90}_{38}\mathrm{Sr} \xrightarrow[\text{28.7 years}]{\text{Half-life}} {}^{90}_{39}\mathrm{Y} + {}^{0}_{-1}\mathrm{e} \qquad \text{(maximum energy 546 keV)}$$

$$^{90}_{39}\mathrm{Y} \xrightarrow[\text{64 hours}]{\text{Half-life}} {}^{90}_{38}\mathrm{Zr} + {}^{0}_{-1}\mathrm{e} \qquad \text{(maximum energy 2.27 Mev)}$$

Strontium-90, a product of nuclear fission, is fabricated in such a way that the low-energy beta-rays are filtered out, thereby allowing the more energetic beta-rays to be used for therapy. In addition, the sources emit bremsstrahlung radiation, the yield and energy being dependent upon the design of the material in which the beta-radiation is absorbed. Ophthalmic applicators contain a strontium-90 compound in refractory form, in rolled silver sheet which is then embedded in spherical silver cups of a radius of 15 mm. Depending on design, the active discs allow radiation from either the front or back of the applicator, the thickness of silver used for screening the low-energy rays being 0.1 mm (figure 2.8). Once constructed, the applicators are palladium-coated for protection against corrosion. Surface applicators of similar construction are also made using strontium-90, this type having 1 mm thick silver plates containing the radio-isotope screened by 0.1 mm silver and protected with a palladium coating. Strontium-90 is also used in the form of tube or nasopharyngeal applicators. In this application the strontium-90 is contained in glass beads which are in a cylindrical stainless steel head 2 cm in length and 1.8 mm in diameter.

2.3.8 Yttrium-90 sources

Yttrium oxide is pressed and sintered and constructed into rod form for use in pituitary ablation. In certain forms the rod is encased in nylon of 2 mm diameter and fitted into a stainless steel screwed socket. As a pure

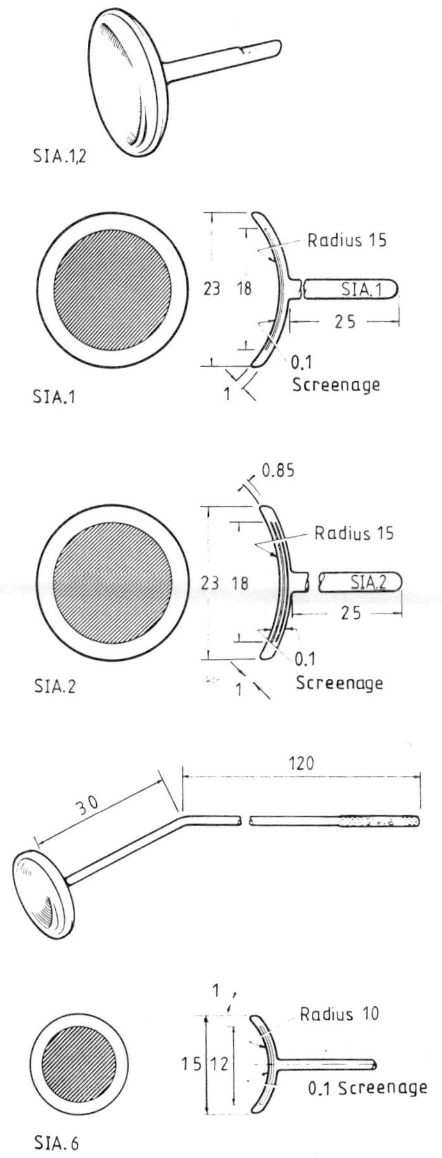

Figure 2.8 Strontium-90 ophthalmic applicators. (Reproduced by kind permission of Amersham International plc.) (Distances given in mm.)

beta-emitter with a half-life of 64 hours and maximum beta energy of 2.27 MeV, activities of up to 300 MBq (8 mCi) per rod are available. The yttrium-90 can be formed by nuclear fission, as a daughter element of $^{90}_{38}Sr$ as shown above or by the (n–γ) reaction

$$^{89}_{39}Y\ (n{-}\gamma)\ ^{90}_{39}Y$$

In practice, the latter process has been used for the production of yttrium rods (Jones *et al* 1963).

2.3.9 Iodine-125 sources

Since 1965 iodine-125 seeds have been used routinely for permanent interstitial implants. The isotope is produced as a daughter element of xenon-125, which is itself produced in an (n–γ) reaction. The activation and decay process for this reaction is

$$^{124}_{54}Xe\ (n{-}\gamma)\ ^{125}_{54}Xe \xrightarrow[\text{8 days}]{\text{Half-life}} {}^{125}_{53}I + {}_{-1}^{0}e$$

Initially, the means of preparing iodine-125 seeds resulted in the presence of iodine-126 contamination due to the neutron activation of iodine-125 itself. By circulating the parent xenon-125 outside the reactor and trapping the iodine-125 shortly after it is formed, significant iodine-126 formation is now avoidable. Iodine-125 has a half-life of 60.1 days and decays by electron capture with the emission of characteristic radiation and Auger electrons. For clinical use the iodine-125 is embedded in two ion-exchange-resin spheres with a radiographic gold marker at the centre (figure 2.9(*a*)). The three spheres are then encapsulated in a welded titanium capsule, 0.05 cm thick, which absorbs the Auger electrons. The principal photon emissions from this isotope are 27.4 and 31.4 keV K_a and K_b characteristic x-rays and a 35.5 keV gamma-ray. This low photon energy means that the half-value layer (HVL) thickness of lead for protection purposes is only 0.025 mm and that the HVL range in tissue is 2 cm.

Recent manufacturing changes include the introduction of iodine-125 seeds in a carrier of a synthetic absorbable suture material and the seeds themselves can be in the form of iodine-125 adsorbed on to a silver rod (figure 2.9(*b*)), the dimensions of the tube remaining at 4.5 mm × 0.8 mm diameter. The benefits of using such a long half-life isotope for implant work include easier scheduling, since they have a half-life of approximately two months. In addition, it is thought that this physical characteristic may result in an improved therapeutic ratio.

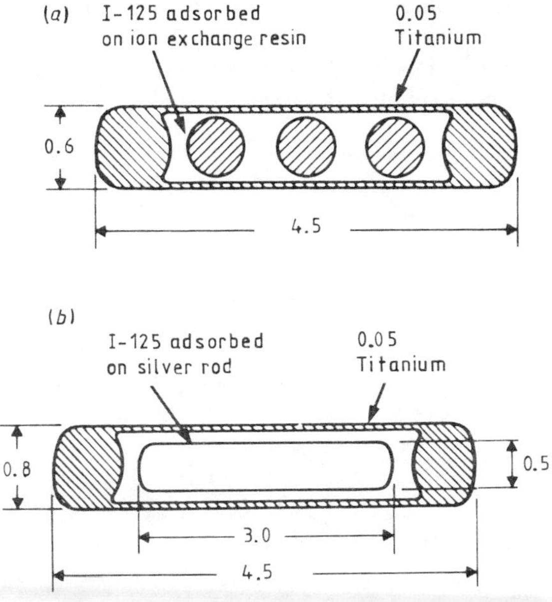

Figure 2.9 Iodine-125 seeds; (*a*) type 6701 and (b) type 6711 (distances given in mm). (Reproduced by kind permission of Amersham International plc.)

2.3.10 Californium-252 sources

Californium-252 is a man-made isotope produced by bombarding plutonium-239 with neutrons in high-flux reactors. The radionuclide decays by alpha-emission with a half-life of 2.64 years and emits neutrons by spontaneous fission with a half-life of 85 years. It is the neutron-emission which makes the nuclide of such interest, but it is the alpha-emission that dictates the practical half-life of a californium-252 source.

The fission spectrum of neutrons has an average energy of 2.1 to 2.3 MeV, with a modal energy of about 1 MeV. The associated x-rays are within the range 0.5 to 1 MeV. The isotope has a high specific activity which makes it a unique type of neutron source, compatible with the high emission rate and small source size required for interstitial implant brachytherapy. Three types of sources are available, tubes, needles and after-loading seeds. The active core is a ceramic mixture of palladium and californium oxide which is placed inside a platinum–iridium alloy tube and drawn out into wire form. The wire cladding and capsule wall combine to give a total filtration of 0.25 mm of 10% iridium–platinum alloy which reduces, but by no means eliminates, the high-energy beta-ray content of californium-252 radiation.

3 Theory of Sealed-source Dosimetry

3.1 Introduction

One of the major developments in sealed-source dosimetry was due to Eve (1906). He introduced a quantity k which related the number of ion pairs, J, produced in a given volume of air per unit time to the mass M of a point source of radium placed at a distance d from the measurement point in a very large volume of air. Neglecting any scatter in air, Eve formulated the relationship

$$J = kMe^{-\mu d}/d^2$$

where the exponential terms accounted for the attenuation of the radiation in air. The factor k became known as Eve's number.

With the introduction of the roentgen in 1928 for measuring the exposure due to x-rays, and its subsequent extension to the measurement of gamma-ray photons in 1937, Eve's number became known as the k factor. This factor related the exposure in roentgen to the activity in mCi and was expressed in units of $\mathrm{R\,mCi^{-1}\,h^{-1}}$ at 1 cm. Initially it was introduced to define the exposure rate due to a radium source, but following the ICRU recommendations that the curie could be used to define the activity of any radionuclide (ICRU 1951) it became possible to define the k factor for other isotopes used in brachytherapy. ICRU (1957) changed the name of this factor to the 'specific gamma-ray emission' and then (ICRU 1962) to the 'specific gamma-ray constant, Γ'. This change was made to draw attention to the fact that this quantity is a constant for any radionuclide.

3.1.1 Exposure rate constant and air kerma rate constant

A more fundamental change to the quantity relating exposure to activity was made by the ICRU in 1971 (ICRU 1971). Prior to this the specific gamma-ray constant and its precursors only considered the gamma-ray emission from a radionuclide and neglected the contribution to the exposure at a point from the characteristic x-rays and internal bremsstrahlung emitted by a gamma-ray source. This was rectified by the introduction of the *exposure rate contant*, Γ_δ, which defined the exposure rate from a point source to include any additional photons which have an energy greater than a certain cut-off energy δ. The exposure rate constant, Γ_δ, was thus defined to be

$$\Gamma_\delta = (d^2/A)(\mathrm{d}X/\mathrm{d}t)_\delta \quad (3.1)$$

where $(\mathrm{d}X/\mathrm{d}t)_\delta$ is the exposure rate due to photons of energy greater than δ (expressed in keV) at a distance d from a point source of activity A. The recommended units for this constant are $\mathrm{R\,m^2\,h^{-1}\,Ci^{-1}}$ (ICRU 1971), or any convenient multiple of these. Units conventionally used have been $\mathrm{R\,h^{-1}\,mCi^{-1}\,cm^2}$ or $\mathrm{mR\,h^{-1}\,mCi^{-1}\,m^2}$.

Note $1.0\ \mathrm{R\,h^{-1}\,mCi^{-1}\,cm^2} = [10^3 \times 1 \times 1 \times (1/100)^2]$
$= 0.1\ \mathrm{mR\,h^{-1}\,mCi^{-1}\,m^2}$.

The value of Γ_δ is derived by considering the number of photons crossing an area of 1 $\mathrm{cm^2}$, the energy fluence, their probability of interacting in air and the average energy lost in each interaction. Since the disintegration of a radionuclide yields a spectrum of photon energies together with a spectrum for the bremsstrahlung radiation, the exposure rate constant for a point source is given by

$$\Gamma_\delta = \Gamma + \Gamma_x$$

where Γ is the specific gamma-ray constant of the radionuclide itself and Γ_x is the specific gamma-ray constant which accounts for all photons of energy greater than δ and of non-nuclear origin (Glasgow and Dillman 1982). A knowledge of the gamma-ray spectrum and other pertinent spectroscopy data enables the values of Γ_δ and Γ to be calculated using the equation

$$\begin{aligned}\Gamma_i &= (3.7 \times 10^7\ \mathrm{s^{-1}\,mCi^{-1}})(3.6 \times 10^3\ \mathrm{s\,h^{-1}})(10^6\ \mathrm{eV\,MeV^{-1}}) \\ &\quad \times (4.803 \times 10^{-10}\ \mathrm{ESU})(1.293 \times 10^{-3}\ \mathrm{g\,cm^{-3}})(4\pi)^{-1} \\ &\quad \times (33.85\ \mathrm{eV})^{-1}(E\ \mathrm{MeV})_i(I, \gamma/\mathrm{decay})_i[(\mu_{en}/\rho)\ \mathrm{cm^2\,g^{-1}}]_i \\ &= 194.5(E\ \mathrm{MeV})_i(I, \gamma/\mathrm{decay})_i[(\mu_{en}/\rho)\ \mathrm{cm^2\,g^{-1}}]_i. \end{aligned} \quad (3.2)$$

In this equation Γ_i is the specific gamma-ray constant for photons of energy E_i, $(I, \gamma/\mathrm{decay})$ is the intensity of the gamma photons per decay and $(\mu_{en}/\rho)_i$ is the mass energy absorption coefficient of the photon.

Values of the specific gamma-ray constant and exposure rate constant for radio-isotopes typically used in brachytherapy are given in table 3.1

in the conventional units of $\mathrm{R\,cm^2\,h^{-1}\,mCi^{-1}}$ and the SI equivalent of $\mathrm{C\,kg^{-1}\,m^2\,s^{-1}\,Bq^{-1}}$.

Note $1.0\ \mathrm{R\,cm^2\,h^{-1}\,mCi^{-1}} = (2.58\times10^{-4})(1/3600)(1/3.7\times10^{7})(1/100)^2$
$= 0.194\times10^{-18}\ \mathrm{C\,kg^{-1}\,m^2\,s^{-1}\,Bq^{-1}}$.

The introduction of SI units in radiation dosimetry led the ICRU (1980) to recommend the use of the *air kerma rate constant* as a replacement for the exposure rate constant on the basis that, as the exposure rate constant is used to determine absorbed dose, it would be more convenient to substitute air kerma for exposure. The air kerma rate constant, $\overset{*}{\Gamma}_\delta$, is therefore defined to be

$$\overset{*}{\Gamma}_\delta = (d^2/A)(\mathrm{d}K_{\mathrm{air}}/\mathrm{d}t)_\delta \tag{3.3}$$

where $(\mathrm{d}K_{\mathrm{air}}/\mathrm{d}t)_\delta$ is the air kerma rate due to photons of energy greater than δ (expressed in keV) at a distance d from a point source of activity A. As with the exposure rate constant all photons emitted by the source are included in deriving this definition. The recommended units are $\mathrm{m^2\,J\,kg^{-1}}$ (ICRU 1980) which become $\mathrm{m^2\,Gy\,Bq^{-1}\,s^{-1}}$ when the special units for absorbed dose and activity are used. More convenient units for the air kerma rate constant are $\mathrm{\mu Gy\,h^{-1}\,MBq^{-1}}$ at 1 m or $\mathrm{mGy\,h^{-1}\,mBq^{-1}}$ at 1 cm, the latter units being numerically greater than the former by a factor of 10.

As with the exposure rate constant, the photons included in this definition comprise gamma-rays, characteristic x-rays and internal bremsstrahlung. The quantity is defined for an ideal point source in vacuum, thus corrections are required in its use to account for any media intervening between the source and the point of measurement which give rise to absorption and scattering. Equivalent values of the air kerma rate constant are given in table 3.1, both in $\mathrm{Gy\,s^{-1}\,Bq^{-1}\,m^2}$ and $\mathrm{\mu Gy\,h^{-1}\,MBq^{-1}\,m^2}$.

Note (i) $1.0\ \mathrm{mR\,h^{-1}\,mCi^{-1}\,m^2} = (2.58\times10^{-7}\times33.85)(1/3600)$
$\times(1/3.7\times10^{7})$
$= 6.56\times10^{-17}\ \mathrm{Gy\,s^{-1}\,Bq^{-1}\,m^2}$

(ii) $1.0\ \mathrm{mR\,h^{-1}\,mCi^{-1}\,m^2} = (2.58\times10^{-7}\times33.85)\times10^{6}\times1$
$\times(1/37)$
$= 2.36\times10^{-1}\ \mathrm{\mu Gy\,h^{-1}\,MBq^{-1}\,m^2}$.

A greater study of the relationship between exposure rate constant and air kerma rate constant can be found in the book by Dutreix *et al* (1982).

Table 3.1 Specific gamma-ray, exposure rate and air kerma rate constants. (Sources of data NCRP (1974), Dutreix *et al* (1982), Glasgow and Dillman (1979, 1982) and NBS (1982).)

Nuclide	Specific gamma-ray constant ($R\,h^{-1}\,mCi^{-1}\,cm^2$)	Specific gamma-ray constant ($C\,kg^{-1}\,s^{-1}\,Bq^{-1}\,m^2$) ($\times 10^{-18}$)	Exposure rate constant ($mR\,h^{-1}\,mCi^{-1}\,m^2$)	Exposure rate constant ($C\,kg^{-1}\,s^{-1}\,Bq^{-1}\,m^2$) ($\times 10^{-18}$)	Air kerma rate constant ($\mu Gy\,h^{-1}\,MBq^{-1}\,m^2$)	Air kerma rate constant ($Gy\,s^{-1}\,Bq^{-1}\,m^2$) ($\times 10^{-17}$)
^{60}Co	13.07	2.536	1.307	2.536	0.308	8.574
^{125}I	0.04–0.05	0.08–0.10	0.145	0.281	0.034	0.951
^{137}Cs	3.23	0.627	0.3275	0.635	0.077	2.148
^{182}Ta	6.71	1.302	0.687	1.333	0.162	4.507
^{192}Ir	4.62	0.896	0.469	0.910	0.111	3.077
^{198}Au	2.32	0.450	0.238	0.462	0.056	1.561
^{222}Rn	9.18	1.781	1.027	1.992	0.242	6.737
^{226}Ra	9.07	1.760	0.825†	1.600	0.195	5.412

† Filtered by 0.5 mm Pt.

3.1.2 Exposure rate and air kerma rate from a point source

The exposure rate in air has been conventionally calculated by rewriting equation (3.1) such that

$$\mathrm{d}X/\mathrm{d}t = \dot{X}_{\mathrm{a}} = A\Gamma_{\delta}/d^2 \tag{3.4}$$

where A is the activity in mCi and Γ_{δ} the exposure rate constant. Similarly, the air kerma rate in air is given by rewriting equation (3.3)

$$\mathrm{d}K/\mathrm{d}t = \dot{K}_{\mathrm{a}} = A\overset{*}{\Gamma}_{\delta}/d^2 \tag{3.5}$$

where A is the activity in MBq and $\overset{*}{\Gamma}_{\delta}$ is the air kerma rate constant.

These equations are the basis of brachytherapy dosimetry. Because of their similarity, dose calculations can readily be converted from exposure to air kerma by a single factor, when the appropriate units are used. From equations (1.8) and (1.10) in §1.2.4 we see that the air kerma in cGy is equal to (0.873 × exposure rate), where the exposure rate is in roentgen. In the following chapters equations are derived for exposure rate, $\dot{X}_{\mathrm{P}}$; they could equally well be derived for the air kerma rate $\dot{K}_{\mathrm{a}}$ by changing the term $(A\Gamma_{\delta})$, where A is the activity in mCi and Γ_{δ} the exposure rate constant, to $(A\overset{*}{\Gamma}_{\delta})$ where activity, A, is now in MBq and $\overset{*}{\Gamma}_{\delta}$ is the air kerma rate constant.

3.2 Exposure Rate and Air Kerma Rate in Air from Sources of Simple Geometry

3.2.1 Unfiltered line source

Many of the sources used in brachytherapy are not point but line sources. The exposure rate at any point P from a line source can be calculated by considering the source to be made up of many point sources. If the source has a total activity A and length L, then the linear activity A' is given by $A = A' \times L$. Consider a segment of the source of length $\mathrm{d}l$, then the exposure rate at point P due to this segment is given by

$$\mathrm{d}\dot{X}_{\mathrm{P}} = A'\Gamma_{\delta}\mathrm{d}l/d^2. \tag{3.6}$$

By integrating over the length of the source, from l_1 to l_2 in figure 3.1, the exposure rate at P from the whole line source is given by

$$\dot{X}_{\mathrm{P}} = \int_{l_1}^{l_2} A'\Gamma_{\delta}\,\mathrm{d}l/d^2. \tag{3.7}$$

Since $d = h \sec\theta$ and $l = h\tan\theta$, then $\mathrm{d}l = h\sec^2\theta\,\mathrm{d}\theta$. Substituting for d,

l, and dl in equation (3.7), we have

$$\dot{X}_{\mathrm{P}} = A'\Gamma_\delta \int_{\theta_1}^{\theta_2} \mathrm{d}\theta/h$$

$$= A'\Gamma_\delta(\theta_2 - \theta_1)/h$$

or

$$\dot{X}_{\mathrm{P}} = A\Gamma_\delta[(\theta_2 - \theta_1)/L]h \tag{3.8}$$

where h is the perpendicular distance from P to the line source in cm and θ is the angle in radians between the perpendicular to the line source from P and the line connecting P and the source element, dl.

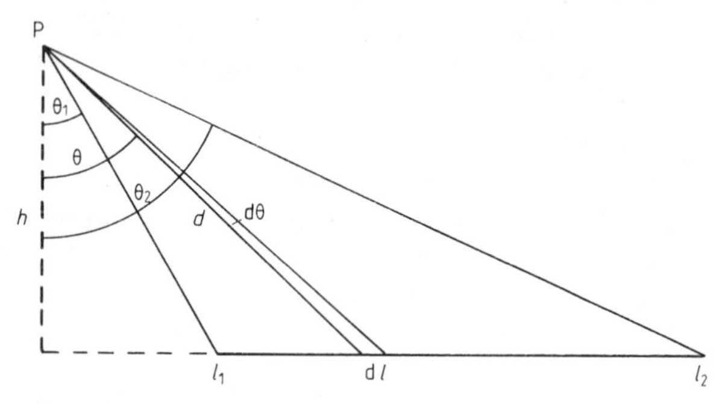

Figure 3.1 Calculation of exposure/air kerma rate around a linear unfiltered radioactive source.

3.2.2 Ring source

To determine the exposure distribution around a uniform circular ring source of radius r and total activity A (figure 3.2), we consider the exposure rate at a point P, due to an element of length dx given by r dθ and linear activity A' a distance d from P. Assuming constant filtration, the exposure rate due to the element is given by

$$\mathrm{d}\dot{X}_{\mathrm{P}} = \Gamma_\delta A' r\, \mathrm{d}\theta/d^2$$

where $d^2 = h^2 + r^2 + y^2 - 2ry \cos\theta$ (see figure 3.2).

The exposure rate, $\dot{X}_{\mathrm{P}}$, due to the whole ring is found by integrating the above equation for θ in the range 0 to 2π

$$\dot{X}_{\mathrm{P}} = A'\Gamma_\delta \int_0^{2\pi} r\, \mathrm{d}\theta/d^2$$

$$= 2\pi A'\Gamma_\delta/[(r^2 + y^2 + h^2)^2 - 4r^2y^2]^{1/2} \tag{3.9}$$

where $2\pi A'$ is the total activity, A, of the ring source.

Using this equation it can be shown that for distances from the plane of the ring greater than $h/r = 0.5$, the isodose lines are parallel to the plane of the ring over the projected area of the ring. This type of distribution is the one obtained for many mould techniques when sources are used in a circle.

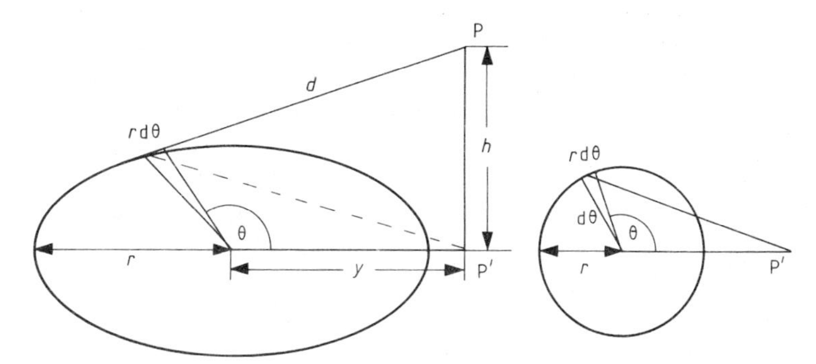

Figure 3.2 Calculation of exposure/air kerma rate at a point due to a ring source of constant filtration.

3.2.3 Circular disc source

The exposure distribution around a uniformly loaded circular disc of total activity A and activity per unit area of A' can be obtained by considering the disc to be made up of a series of rings. The exposure rate, $d\dot{X}_P$, due to one such ring of annular thickness dx and radius x is given by equation (3.9), that is

$$d\dot{X}_P = 2\pi A'\Gamma_\delta x\, dx/[(x^2+y^2+h^2)^2 - 4x^2y^2]^{1/2}. \tag{3.10}$$

Thus the exposure rate due to the whole disc, of radius r, is found by integrating equation (3.10) over the radius of the disc, that is

$$\begin{aligned}\dot{X}_P &= \Gamma_\delta 2\pi A' \int_0^r x\, dx/[(x^2+y^2+h^2)^2 - 4x^2y^2]^{1/2} \\ &= \pi A'\Gamma_\delta \ln[\![\{r^2 - y^2 + h^2 + [(r^2-y^2+h^2)^2 + 4y^2h^2]^{1/2}\}/2h^2]\!].\end{aligned} \tag{3.11}$$

Since the total activity of the disc is A, then the term $\pi A'$ can be substituted by A/r^2 because $A' = A/\pi r^2$. For points on the central axis, where $y = 0$, equation (3.11) reduces to

$$\dot{X}_P = A\Gamma_\delta \ln(1 + r^2/h^2)/r^2.$$

The isodose distribution around a disc is elliptical in form.

3.2.4 Cylindrical source

These sources can be either in the form of a cylindrical tube, when it is advantageous to know the dose within the cylinder as well as outside, or as a solid cylinder of radioactive material. Spiers (1947) has discussed the dose distribution of the former and obtained an expression for the intensity at a point due to the whole cylinder by integrating equation (3.9), for a ring, over the whole length of the cylinder. To evaluate this integral 'elliptical integral' tables have to be used. In the case of a solid cylinder Parker (1947) derived an expression for the intensity at any point by integrating equation (3.11) for a disc source over the length of the tube.

3.2.5 Spherical sources

Using the integration approach, Souttar (1931) and Mayneord (1945) showed that the exposure rate at a distance d from the centre of a sphere of radius R, containing a uniformly distributed radionuclide of activity A throughout the volume, is given by

$$\dot{X}_P = 3\Gamma_\delta(A/4)R^3\{2R + [(R^2 - d^2)/d]\ \ln[(R + d)/(R - d)]\}. \qquad (3.12)$$

Exposure distributions around sources of simple geometry have been discussed in more detail by Mould (1966).

3.3 Filtration and Self-absorption

The approach used to derive the exposure from sources of simple geometry assumes a 'bare' source in air. However, sealed sources are encapsulated and thus consideration must be given to the absorption of radiation by the source capsule and self-absorption within the source. The attenuation due to these processes results in a reduction in the exposure rate or air kerma rate at the point of interest by a factor of $f_1(d, \theta)$.

3.3.1 Filtration of line source

For a point source encapsulated in a capsule of constant thickness, the attenuation factor, $f_1(d, \theta)$, is constant. With a uniformly filtered line source, however, there is a variation in this factor due to rays emanating from different points in the source being attenuated by differing amounts depending on the relative position of the origin of the ray with respect to the point of calculation. The degree of variation is dependent upon the different thickness of tube, T, traversed by the rays (figure 3.3). In this situation the exposure rate at P is given

$$\dot{X}_P = (A'\Gamma_\delta/h) \int_{\theta_1}^{\theta_2} \exp(-\mu_{en} T \sec\theta)\, d\theta$$

$$= (A'\Gamma_\delta/h) \left[\int_0^{\theta_2} \exp(-\mu_{en} T \sec\theta)\, d\theta - \int_0^{\theta_1} \exp(-\mu_{en} T \sec\theta)\, d\theta\right] \quad (3.13)$$

where μ_{en} is the energy absorption coefficient per unit length of filtering material for the photons from the particular radionuclide.

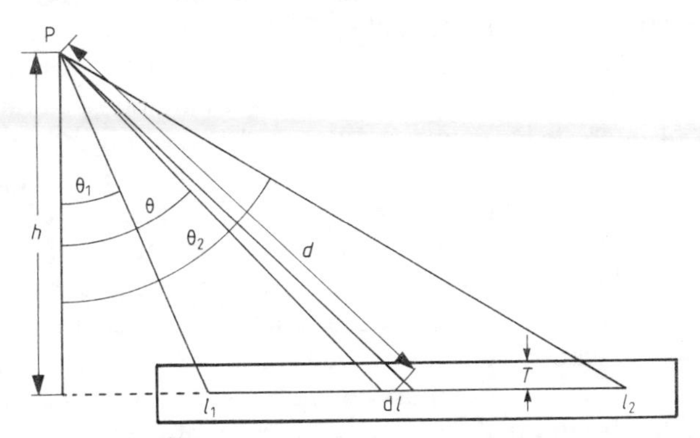

Figure 3.3 Calculation of exposure/air kerma rate around a linear filtered radioactive source where the thickness of filtration is T.

Another approach to computing the dose from a filtered linear source is the interval method. This method divides the source into a large number of point sources and sums the contribution each segment of the source makes to the point of calculation. It has been used by Shalek and Stovall (1968) to compute the exposure around a radium tube and by Breitman (1974) for calculating the absorbed dose distribution for caesium tubes.

In both methods the value chosen in the calculation for the wall thickness, T, must be the effective wall thickness, to account for the fact that since the internal diameter of the source is of finite size, some of the gamma-rays originating from the source axis pass through the filter in a non-radial direction. Using the method of calculation suggested by Evans and Evans (1948), Keyser (1951) calculated the effective wall thickness and his results indicate that for a typical wall thickness of 0.5 mm, the effective wall thickness is 0.58 mm. This is the value of T to be used in equation (3.13). The value of μ_{en} has also to be chosen with care since for radionuclides with a complex gamma-ray spectrum, one cannot assume that the transmission of gamma-rays through the filter decreases exponentially with the thickness of the filter, thus the absorption coefficient varies as the gamma-rays traverse the filter. In the case of radium, Whyte (1955) determined by measurement the transmission of the radium gamma-rays through platinum. From these measurements 'the effective absorption coefficient', defined as an absorption coefficient which would produce the observed decrease in gamma-ray intensities assuming exponential absorption, can be determined. This value is used in the Sievert integration technique. However, in the interval technique different effective absorption coefficients are used for the different thicknesses of platinum through which the gamma-rays pass in reaching the point of interest, P. Young and Batho (1964) compared the effect of assuming a single value for μ_{en} of 15.2 mm^{-1} for points opposite the active length of a radium source with radial wall thickness of 0.5 mm of platinum with those determined by decreasing the value of μ_{en} with increasing thickness of absorber. The results showed variations of less than 2% over most points around the source. However, for points close to the axis but remote from the active length, the divergence was 5% or more. In conclusion, the authors commented that 'no useful analytical expression has been found for μ_{en} as a function of T which would permit integration of the integral required for a linear filtered radium source'. However, Shalek and Stovall (1968), attempting a different method of calculating the effective absorption coefficient, suggested a value of 17.0 mm^{-1} for 0.5 mm and 15.0 mm^{-1} for 1 mm of platinum for photons from radium.

Another approach to describing the absorption in platinum from radium gamma-rays has been considered by Toepfer and Rosenow (1980). Using the data of Keyser (1951) and Whyte (1955), they proposed an elementary function of the form

$$T^{-1} = (1 + \mu_0 d) T_0^{-1}$$

to describe the fractional transmission T of radium gamma-rays through a platinum absorber of thickness d. (Note: the term 'T' is not the same as that used to denote wall thickness in the previous cases.) Using a least square fit technique on the data, values for the constants T_0 and μ_0 were determined such that $T_0 = 0.9780$ and $\mu_0 = 1.501\ \mathrm{cm}^{-1}$. A comparison between transmission factors calculated with these parameters and the experimental data showed a maximum difference of 0.4% if the minimum value of $d = 0.0148$ cm was excluded.

This work also led Toepfer and Rosenow to formulate an expression describing the variation of the specific gamma-ray constant of radium with platinum filtration of the form

$$\Gamma(d) = \Gamma_0/(1 + \mu_0 d)$$

where $\mu_0 = 1.501\ \mathrm{cm}^{-1}$ and $\Gamma_0 = 6.356 \times 10^{-5}\ \mathrm{A\,m^2\,kg^{-2}}$ ($8.869\ \mathrm{R\,cm^2\,h^{-1}\,mg^{-1}}$), which is based on obtaining the recommended value of $8.25\ \mathrm{R\,cm^2\,h^{-1}\,mg^{-1}}$ for 0.5 mm of platinum. Whilst this problem of selecting the correct value of Γ and μ_{en} is real for radium calculations, the results are less critical for caesium and for sources which have a thin sheath of platinum for filtration.

3.3.2 Sievert integral

In calculating the exposure from a filtered linear source it is necessary to evaluate the integrals in equation (3.13). Sievert (1932) published extensive tables of the integral

$$U(\theta_i) = \int_0^{\theta_i} \exp(-\mu_{en} T \sec\theta)\ \mathrm{d}\theta \tag{3.14}$$

for values of θ between $\pi/6$ and $\pi/2$. These tables have been extended by Shalek and Stovall (1969) by numerical integration according to Simpson's rule, and the values derived are presented in table 3.2.

Using these tables it is possible to manually calculate the exposure at any point around a line source. For easier computation Young and Batho (1964) have expressed the Sievert integral as a series, whilst Busch (1968) represented the integral by a polynomial. Williamson *et al* (1983b) carried out a Monte Carlo evaluation of the Sievert integral for radium-226 and platinum-clad iridium-192 seeds and found that their results agreed with those obtained using the Sievert integral to within ±2% in the region bounded by the active source ends, but the disagreement was greater outside this region.

Table 3.2 Table of Sievert integrals. (Abridged data taken from Shalek and Stovall (1969).)

		$\int_0^\theta \exp(-\mu_{en}T/\cos\theta)\,d\theta$ for various values of $\mu_{en}T$					
Angle θ	Tan θ	$\mu_{en}T = 0.00$	0.05	0.10	0.15	0.20	0.25
4	0.070	0.070	0.066	0.063	0.060	0.057	0.054
8	0.141	0.140	0.133	0.126	0.120	0.114	0.109
12	0.213	0.209	0.199	0.189	0.180	0.171	0.163
16	0.287	0.279	0.266	0.252	0.240	0.228	0.217
20	0.364	0.349	0.332	0.315	0.300	0.285	0.270
24	0.445	0.419	0.398	0.378	0.359	0.341	0.324
28	0.532	0.489	0.464	0.440	0.418	0.397	0.377
32	0.625	0.559	0.530	0.503	0.477	0.452	0.429
36	0.727	0.628	0.596	0.556	0.535	0.507	0.481
40	0.839	0.698	0.661	0.626	0.593	0.561	0.531
44	0.966	0.768	0.726	0.687	0.650	0.615	0.581
48	1.111	0.838	0.791	0.747	0.706	0.667	0.630
52	1.280	0.908	0.856	0.831	0.761	0.718	0.677
56	1.483	0.977	0.920	0.866	0.815	0.768	0.723
60	1.732	1.047	0.983	0.924	0.868	0.816	0.766
64	2.050	1.117	1.046	0.982	0.919	0.861	0.807
68	2.475	1.187	1.108	1.035	0.967	0.904	0.845
72	3.078	1.257	1.168	1.087	1.012	0.943	0.879
76	4.011	1.326	1.226	1.135	1.052	0.976	0.907
80	5.671	1.396	1.281	1.178	1.086	1.003	0.928
84	9.514	1.466	1.330	1.212	1.110	1.019	0.939
88	14.301	1.536	1.362	1.228	1.118	1.024	0.942
90	∞	1.571	1.365	1.229	1.118	1.024	0.942

3.3.3 Self-absorption

In calculating the dose from a source, correction has to be made to allow for the gamma-rays which interact within the source itself and thus are internally attenuated. Measurements of this effect show variations of from 1 to 10% self-absorption. The theoretical basis of this process has been discussed by Evans and Evans (1948), Evans (1955), Trucco (1962) and Morris (1964) and various approximations have been devised to account for this effect (Evans 1968). For practical purposes in brachytherapy, where the internal diameter of the source is only about 1 mm, the correction is of the order of 0.5% (Shalek and Stovall 1968). Young and Batho (1964) considered this problem for radium tubes when considering the effective wall thickness of the radium container and derived an equation for the effective wall thickness which incorporated both self-absorption and wall filtration such that

$$d = (2/\pi R^2) \int_{-R}^{+R} (R^2 - v^2)^{1/2} [(R+t)^2 - v^2]^{1/2} \, dv - 8R/3\pi[1 - (\mu'/\mu)] .$$

In this expression μ' is the effective absorption coefficient of the radium salt ($\mu' = 0.1\mu$), t is the radial thickness of the platinum, R is the radius of the salt, μ is the absorption coefficient in the platinum sheath and (u, v) is the position of a point in the source from which the photon emanates.

Using this value of the effective wall thickness d, the authors calculated their tables relating the exposure rate at any point to points outside the source.

3.4 Photon Attenuation and Scattering in a Medium

Until the advent of artificially produced isotopes the assumption was made that, to a first approximation, the attenuation of photons in the vicinity of an implanted source was balanced by a corresponding increase in the number of scattered photons produced. Brachytherapy dosimetry was therefore based for more than 50 years on the assumption that the exposure dose at a point in tissue for a radium source was the same as the exposure dose in air at the same point. In the early 1950s the validity of this assumption began to be studied and experimental determinations were made of the absorption and scattering in water for the gamma-rays (a) for radium-226 by Hine and Freidman (1950) and Ter-Pogossian *et al* (1952), (b) for radium-226 and cobalt-60 by Wootton *et al* (1954) and Reuss and Brunner (1957) and (c) for radium-226, cobalt-60, gold-198 and mercury-203 by Van Dilla and Hine (1952).

These experiments showed that the ratio of the exposure in water to the exposure in air at a given distance from the source varied from unity, and in general decreased with increasing distance from the source. This ratio can be termed the tissue–air ratio (TAR) for brachytherapy dosimetry and is represented by the factor $f_2(d, \theta)$ below. As other radium substitutes became available further measurements were made of the ratio. These were carried out (a) by Kartha *et al* (1966) for radium-226, cobalt-60 and caesium-137, (b) by Meredith *et al* (1966) for radium-226, cobalt-60, caesium-137, gold-198, iridium-192 and tantalum-182, (c) by Smocovitis *et al* (1967) for radium-226 and (d) by Meisberger *et al* (1968) for gold-198, iridium-192, caesium-137, radium-226 and cobalt-60. A comparison of these results carried out by Meisberger *et al* (1968) showed agreement within about 2.5% over the range 0 to 10 cm with the exception of the data of Wootton *et al* (1954) which were several per cent higher for distances of 5 to 10 cm. Over the years numerous methods have been adopted to derive numerical expressions to define the brachytherapy TAR, $f_2(d, \theta)$.

3.4.1 Build-up factor

Evans (1955, 1968) formulated equations to describe the absorption and scattering of photons within a medium in terms of an exponential function to account for the primary photon attenuation and a build-up factor to describe the complicated contribution of scattered and secondary photons to exposure. This factor, B, was defined as

$$B = \text{(total observed dose rate)}/\text{(primary dose rate)}.$$

To a reasonable approximation, Evans suggested that the increase in the build-up factor, B, with distance from a point source in a medium could be represented by

$$B = 1 + a(\mu d)^K \qquad (3.15)$$

where a and K are constants, μd is the distance, d, from the source measured in mean free paths of the primary radiation.

The relative magnitude of these two effects, absorption and scattering, was studied by Hale (1958) who found that for gamma-ray energies of approximately 0.4 MeV attenuation balances scatter, whereas for energies less than 0.4 MeV scatter dominates and above this value absorption dominates. Hale therefore proposed that the dose in water could be related to the dose in air by the equation

$$D_w = D_a(AB)$$

where D_w is the dose in water, D_a the dose in air, A the absorption factor (typically an exponential function) and B the build-up factor.

The product (AB) is the ratio of exposure in water to exposure in air and thus defines the factor $f_2(d, \theta)$. Using theoretical data for the penetration of gamma-rays in water, Hale formulated an equation for this factor such that

$$(AB) = f_2(d, \theta) = \exp(-\mu d)\exp[(0.77/E^{0.29})\mu d]$$

for $0.255 < E < 2$ MeV and $0 < \mu d < 1$.

A similar relationship was derived by Kartha *et al* (1966) following measurements made on radium-226, caesium-137 and cobalt-60. The expression they derived is of the form

$$f_2(d,\theta) = \exp\{[(0.73/E^{0.05}) - 1]\mu d\}.$$

Tables of energy absorption build-up factors, $B_{en}(\mu_{tot}d)$, were produced by Berger (1968), the factor being given at a distance d in the units of mean free path, $(\mu_{tot}d)$, where μ_{tot} is the total attenuation coefficient for the medium. These tables have been used as the basis for numerous calculation techniques in sealed-source brachytherapy. Tripathi and Shanta (1985) have made a graphical and analytical study of these data and derived relationships between $B_{en}(\mu_{tot}d)$, E and $\mu_{tot}d$ which they then used to derive a TAR for point isotropic sources. This is of the form

$$\mathrm{TAR}(d) = aE_{av}^{b}$$

where $a = a_1 + a_2d$ and $b = b_1 + b_2d$. The derived values for the parameters a_1, a_2, b_1 and b_2 have been chosen so that they are the same for all radionuclides commonly used in brachytherapy and the authors claim that the generalised relationship gives the value of the build-up factor at any depth up to 30 cm to within $\pm 2\%$.

3.4.2 Empirical functions

Using the data of Wootton *et al* (1954), Batho and Young (1964) derived an equation for $f_2(d,\theta)$ of the form

$$f_2(d,\theta) = 1 - md^n.$$

In the light of the results of Smocovitis *et al* (1967), Batho and Young (1967) chose values of m and n such that

$$f_2(d,\theta) = 1 - 0.010d^{5/4}. \qquad (3.16)$$

Another approach was adopted by Meisberger *et al* (1968) who, on studying the experimental data of numerous authors, including their own, derived a third-order polynomial to define the ratio of exposure in water to exposure in air. This was of the form

$$f_2(d,\theta) = A + Bd + Cd^2 + Dd^3. \qquad (3.17)$$

This expression is valid for values of d between 1 and 10 cm, where d is the path length of the radiation in water. The values of the constants A, B, C and D for various radionuclides are given in table 3.3. The Meisberger polynomial is the most commonly applied tissue correction factor.

Table 3.3 Values of A, B, C, D for various radionuclides.

Radionuclide	A	B	C	D
Cobalt-60	9.9423×10^{-1}	-5.318×10^{-3}	-2.610×10^{-3}	1.327×10^{-4}
Caesium-137	1.0091×10^{0}	-9.015×10^{-3}	-3.459×10^{-4}	-2.817×10^{-5}
Iridium-192	1.0128×10^{0}	5.019×10^{-3}	-1.178×10^{-3}	-2.008×10^{-5}
Gold-198	1.0306×10^{0}	-8.134×10^{-3}	1.111×10^{-3}	-1.597×10^{-4}
Radium-226	1.0005×10^{0}	-4.423×10^{-3}	-1.707×10^{-3}	7.448×10^{-5}

To make computational techniques easier, Van Kleffens and Star (1979) proposed an alternative relationship for $f_2(d,\theta)$:

$$f_2(d,\theta) = (1 + ad^2)/(1 + bd^2) \tag{3.18}$$

where a and b are given in table 3.4 for various radionuclides. Comparison of expression (3.18) for caesium-137 gamma-rays with the Meisberger relationship showed differences of less than 1% for d less than 9 cm. For d greater than 9 cm the Meisberger expression drops sharply to 0.465 at 20 cm, whilst the Van Kleffens and Star relation gives a more realistic value of 0.812 for $d = 20$ cm.

Table 3.4 Values of a and b for various radionuclides.

Radionuclide	a	b
Caesium-137	0.0083	0.0108
Cobalt-60	0.0100	0.0145
Radium-226	0.0068	0.0097

3.4.3 Monte Carlo approach

Studies of dose distributions around point-source gamma-emitters have been carried out by Webb and Fox (1979) using Monte Carlo methods for cobalt-60, caesium-137, iridium-192, gold-198 and radium-226, the results obtained being comparable with measured data. Using these data Kornelsen and Young (1981) showed that the results could be well represented by equation (3.15) for cobalt-60 and caesium-137, but the fit was not quite so good for nuclides with complex and/or low-energy spectra, although still very acceptable.

Monte Carlo techniques were also used by Masterson *et al* (1979) to calculate the exposure rate constant and the absorbed dose in water per unit activity at various distances from a stainless-steel-encapsulated iridium-192 seed. By taking the ratio of the absorbed dose rate per unit activity and the exposure rate constant, and correcting to convert absorbed dose into exposure using the factor 1/0.96, they calculated values of $f_2(d, \theta)$ which are in agreement with the measured data of Meredith *et al* (1966). The use of the $f_2(d, \theta)$ factor for iridium-192 has been recently reviewed by Glasgow (1981) and this had led Anderson (1983) to suggest that in view of the large spread in experimental data, $\pm 4\%$, it is justifiable for practical dose calculations to obtain the dose in tissue from the product of exposure in air and a constant absorbed dose to exposure factor of 0.96 cGy R^{-1}.

A more detailed study of the absorbed dose within tissue surrounding nuclides used in brachytherapy has been made by Dale (1982, 1983) using the Monte Carlo calculation technique. This analysis not only extended the range of nuclides to cover iodine-125 and caesium-131 but also provided a method of calculating the absorbed dose within the different body tissues, such as adipose tissue body fat and connective tissue. Calculating a radial dose function, which is essentially a modifying expression to correct the inverse-square fall-off around a point source, Dale derived either a second- or third-order polynomial, similar in form to equation (3.17), to account for all the scattering and attenuation effects occurring in the given tissue (see §3.5). The results indicate that the polynomials for water can be reliably used to describe the fall-off in other tissues, with the exception of low-energy-emitting nuclides where doses in body tissue are significantly different from those in water.

3.5 Absorbed Dose Calculations in a Medium

The absorbed dose rate in a medium due to a source can therefore be evaluated using the equation

$$\dot{D}_P(d, \theta) = [A\Gamma_\delta f_1(d, \theta) f_3(d, \theta)] f_2(d, \theta)/d^2. \qquad (3.19)$$

The terms $f_1(d, \theta)$ and $f_2(d, \theta)$ are the factors utilised in the previous sections to describe filtration, self-absorption and attenuation and scattering in a medium and $f_3(d, \theta)$ is the factor to convert exposure or air kerma to absorbed dose in tissue (§1.2.4). The term in the square brackets in equation (3.19) is often referred to as the source strength, S, quoted in units of cGy h^{-1} at 1 cm.

Over the years many authors have used different methods of evaluating equation (3.19) to calculate the absorbed dose due to sealed sources of different geometry. The use of the Sievert integral with a tissue correction factor, and the 'build-up' factor method proposed by Berger (1968) do not make any correction for attenuation along the length of the source. The interval method adopted by Shalek and Stovall (1968) does, however, make some correction for this but, like the other methods, it assumes that the source has no finite diameter which therefore necessitates the use of 'effective' filter thickness and/or an 'effective' attenuation coefficient. A more fundamental approach to brachytherapy dose calculation was suggested by Milan (1975)—the quantisation method, which essentially extends the interval method to three dimensions. Initially, this method was time-consuming in calculation but Cassell (1983) has developed algorithms which substantially reduce computational time.

A semi-empirical model for dose calculation which separates the scattering and absorption characteristics of the emitted radiation from the dominant geometric factors has been suggested by Mohan and Anderson (1974). This separation is analogous to removing the distance dependence from percentage depth dose to generate tissue air ratios or tissue phantom ratios. We know from equation (3.19) that the absorbed dose rate in air for a point source can be written as

$$\dot{D}_0 = S/d^2 \quad \text{cGy h}^{-1} \tag{3.20}$$

where S is the source strength specified in cGy cm^2 h^{-1}, as recommended by NCRP (1974). Similarly, for line sources, in the absence of a scattering or absorption medium, we know from equation (3.8) that the dose rate is given by

$$\dot{D}_0 = S(\theta_2 - \theta_1)/Lh \quad \text{cGy h}^{-1} \tag{3.21}$$

where L is the active length.

When these sources are placed within a medium, the dose rate $\dot{D}_P$ at P includes the geometric attenuation as well as scatter and attenuation in tissue. By taking the quotient of $\dot{D}_P$ and the corresponding dose 'in vacuum', $\dot{D}_0$, Mohan (1981) defined the *relative dose rate factor*, R, to be

$$R = \dot{D}_P/\dot{D}_0. \tag{3.22}$$

The relative dose rate factor, a dimensionless quantity, is a function slowly varying with position. This means that linear interpolation within a relatively small table of data can be carried out without significant error. Tables of relative dose rate factors are one-dimensional (functions of d) for point sources and two-dimensional (functions of x and h) for line sources.

Using these expressions and introducing a geometric attenuation factor G, which for point sources is $1/d^2$ and for line sources is $(\theta_2 - \theta_1)/Lh$, the dose rate within a medium can be written as

$$\dot{D}_P = SRG. \tag{3.23}$$

Equation (3.23) can be used to express source strength in terms of activity instead of $\mathrm{cGy\,cm^2\,h^{-1}}$ since

$$\dot{D}_P = A\Gamma_\delta f_3(d, \theta)RG = AF_DG$$

where $F_D = (\Gamma_\delta f_3(d, \theta)R)$ and is called the *specific dose rate factor* (Anderson 1975). From equations (3.22) and (3.23) one can see that the relative dose rate factor, R, and the specific dose rate factor, F_D, are related by the expression

$$F_D = (S/A)R.$$

A similar approach to defining the variation in dose has been suggested by Dale (1982, 1983). In this work the dose to an infinitely thin shell of radius r is derived from the equation

$$D(r) = (1/4\pi r^2) \sum_{i=1}^{N} [(\mu_{en}/\rho)_{E_i} E_i \sec \theta_i]$$

where the summation is taken over the whole spectrum of photon energies, E_i, at radius r.

This equation can be used to evaluate the dose rate at 1 cm from a point source of unit activity, which is referred to as the *specific dose constant* (SDC) and written as D_s, with units of $\mathrm{Gy\,m^2\,s^{-1}\,Bq^{-1}}$. Using the SDC as a normalising factor, Dale defined a further factor, the *radial dose factor* (RDF), $g(r)$, such that

$$g(r) = (D(r)/D(1))r^2$$

where $D(1)$ is the dose rate at 1 cm. Thus the absorbed dose rate from a point source of activity A is given by

$$\dot{D}_r = AD_s g(r)/r^2.$$

Using this approach, Dale has derived values for the SDC and the RDF for various tissue/nuclide combinations and then described the RDF,

Table 3.5 Table of specific dose constants (from Dale (1982, 1983)). (Units: $Gy\,m^2\,s^{-1}\,Bq^{-1} \times 10^{-17}$. To convert to $cGy\,cm^2\,h^{-1}\,mCi^{-1}$ multiply figures by 1.332.)

	Specific dose constants Tissues and their densities in $g\,cm^{-3}$						
Nuclides	Water 1.00	Adipose tissue 0.92	Body fat 0.92	Connective tissue 1.20	Prostate 1.05	Rectum 1.04	Tongue 1.04
Cobalt-60	9.24	9.33	9.36	9.09	9.15	8.86	9.22
Iodine-125	1.01	0.67	0.55	1.01	1.02	1.02	0.96
Caesium-137	2.32	2.33	2.34	2.28	2.29	2.22	2.30
Iridium-192	3.35	3.37	3.38	3.34	3.30	3.24	3.36
Gold-198	1.69	1.70	1.70	1.67	1.67	1.61	1.68
Radium-226	6.62	6.66	6.65	6.64	6.51	6.31	6.56

$g(r)$, by a second- or third-order polynomial function of the form

$$g(r) = A_0 + A_1 r + A_2 r^2 + A_3 r^3.$$

From the calculated data, values for the polynomial coefficients A_0, A_1, A_2 and A_3 for various brachytherapy nuclides in water and various tissues have been obtained using least-square curve-fitting tech-

Table 3.6 Polynomial coefficients of $g(r)$ for nuclides in water and various tissues. (From Dale (1982, 1983, 1986).)

Tissue	Nuclide	A_0	A_1	A_2	A_3
Water	Cobalt-60	1.0149	-0.1233×10^{-1}	-0.3229×10^{-3}	–
	Iodine-125	1.1724	-0.1726×10^{0}	0.6487×10^{-2}	–
	Caesium-137	1.0005	-0.2173×10^{-2}	-0.9003×10^{-3}	–
	Iridium-192	0.9840	0.1814×10^{-1}	-0.1863×10^{-2}	–
	Gold-198	0.9887	0.1290×10^{-1}	-0.1674×10^{-2}	–
	Radium-226	1.0056	-0.4958×10^{-2}	-0.6444×10^{-3}	–
Connective	Cobalt-60	1.0300	-0.2068×10^{-1}	–	–
	Iodine-125	1.2497	-0.2656×10^{0}	0.1957×10^{-1}	-0.4924×10^{-3}
	Caesium-137	1.0036	-0.5833×10^{-2}	-0.9048×10^{-3}	–
	Iridium-192	0.9834	0.1754×10^{-1}	-0.2205×10^{-2}	–
	Gold-198	0.9942	0.1275×10^{-1}	-0.1949×10^{-2}	–
	Radium-226	1.0083	-0.4196×10^{-2}	-0.9688×10^{-3}	–
Adipose	Iodine-125	1.0028	0.1593×10^{-1}	-0.1466×10^{-1}	0.6471×10^{-3}
Body fat	Iodine-125	0.9732	0.7323×10^{-1}	-0.1995×10^{-1}	0.7828×10^{-3}

niques which have a correlation coefficient of 0.995 or greater. Values of the specific dose constants and the polynomial coefficients for the radial dose factors for various body tissues are given in tables 3.5 and 3.6.

3.6 Practical Dose Distributions in a Medium for Point and Linear Sources

3.6.1 Point sources

3.6.1.1 Caesium-137. Certain spherical sources are constructed with a constant wall thickness (see figure 2.4) and can be treated as point sources provided account is taken of the attenuation of the photons within the source material and the wall. The dose distribution around such sources is then obtained by evaluating equation (3.19). However, with certain constructions the spherical beads of caesium are encased within cylindrical shells (figure 2.5). Diffey and Klevenhagen (1975) considered the absorbed dose rate due to such a source, type CDCK, using the interval method. This type of source consists of a single caesium bead within a cylindrical shell and they showed that the dose rate at a point $P(d, \theta)$ could be approximated by the expression

$$\dot{D} = (\Gamma_\delta f_3(d,\theta) A f_1(d,\theta)/2L) \int_{-L}^{L} [\exp(-\mu_1 x_1 - \mu_2 x_2)]\,[f_2(d,\theta)]\, d^{-2}\,\mathrm{d}L \tag{3.24}$$

where μ_1 is the energy absorption coefficient for the source material, x_1 is the path length of radiation in the source material, μ_2 is the effective attenuation coefficient for the capsule, x_2 is the path length of radiation in the capsule, $2L$ is the active length of the seed and d is the distance from the centre of the seed to point P. The terms $\Gamma_\delta, f_1(d,\theta), f_2(d,\theta)$ and $f_3(d,\theta)$ are as defined previously.

They found that even though they did not integrate over the total volume of the source, the values of dose rate obtained were in no better agreement with experimental dose rates than those obtained using the method of Young and Batho (1964) to determine the effective wall thickness of the source. Using this computational model the radiation distribution around a CDCK cylindrical bead was shown to be within 3% of the measured dose rate for any point in water. In order to derive the true activity content of the source within the capsule, the authors measured the exposure rate in air, $\dot{X}_c$, at a defined distance from the bead, d_c, and determined the equivalent activity, A_e, such that

$$A_e\Gamma_\delta = \dot{X}_c d_c^2.$$

Using this value for A_e the true activity was obtained from

$$A = A_e g$$

where g is equivalent to $1/f_1(d,\theta)$. The authors showed that the value of g is given by

$$g = \left\{(d_c^2/2L)\int_{-L}^{L}[\exp(-\mu_1 x_1 - \mu_2 x_2)]/d^2\,\mathrm{d}L\right\}^{-1}. \qquad (3.25)$$

Since the calibration distance d_c is large compared with the source dimensions, the expresion for g reduces to

$$g = \exp(\mu_1 L + \mu_2 t) \qquad (3.26)$$

where t is the wall thickness of the capsule.

In these sources the active material, caesium chloride, is embedded in a 2 mm diameter spherical glass bead which is then encased in stainless steel of 0.5 mm thickness. To effect the calibration it is necessary to know the gamma-ray attenuation coefficient of the source capsule. According to Hubbell (1969), the energy absorption coefficient, μ_{en}, and total attenuation coefficient, μ, for stainless steel at 0.662 MeV are 0.217 and 0.566 cm^{-1} respectively. Since the slope of the transmission curve, which is proportional to the attenuation coefficient, increases with increasing wall thickness, Diffey and Klevenhagen measured an effective attenuation coefficient, μ_2, of 0.34 cm^{-1} in water, which is not the same as the narrow beam energy absorption coefficient for stainless steel. The value chosen for μ_1 was 0.085 cm^{-1}, which is the average of the energy absorption coefficients at 0.662 MeV for caesium chloride (0.09 cm^{-1}) and glass (0.075–0.084 cm^{-1})(Hubbell 1969). Using these data in equation (3.26), a value of 1.026 was calculated for g.

3.6.1.2 Iodine-125. The dosimetry of iodine-125 seeds has been studied extensively over the past decade. Due to their construction (see §2.3.9), the seeds show marked anisotropy in the distribution of radiation (Ling *et al* 1983b), although this is generally ignored in dose computations since it is not always practicable to define the orientation of the seeds. Krishnaswamy (1978) calculated the distributions around such seeds (type 6701) using the Berger (1968) build-up equation

$$\dot{D}(r) = (KNE/100 \times 4\pi r^2)\mu_{en}[\exp(-\mu_{tot}r)]B(\mu_{tot}r) \qquad (3.27)$$

where $\dot{D}(r)$ is the dose rate in cGy mCi^{-1} h^{-1}, N is the number of photons emitted per hour per mCi source strength, E is the mean photon energy in keV, μ_{en} and μ_{tot} are the energy absorption and total absorption coefficients in tissue, as quoted by Hubbell (1969), $B(\mu_{tot}r)$ is the energy absorption build-up factor for a distance r from the source in units of mean free path ($\mu_{tot}r$) and K is the conversion factor from keV to ergs.

Calculations of the dose rate were made to points in a grid for distances 5 cm along the transverse axis and 5 cm along the source axis of the seed and a table produced detailing the cGy per hour from a millicurie-equivalent source. In the region of small solid angle subtended by the gold marker, interpolated results were used. The calculated depth doses were within 5% of those measured and these data have been used by Mohan and Anderson (1978) as the basis of the specific dose rate factor for iodine-125. Krishnaswamy (1978) measured an exposure rate of

Table 3.7 Specific dose rate and relative dose rate factors for iodine-125 seeds.

Distance (cm)	Specific dose rate factor† (cGy cm^2 mCi^{-1} h^{-1})	Relative dose distribution factor‡
0.057	1.10	—
0.5	1.11	1.05
1.0	1.10	1.00
1.5	1.04	0.92
2.0	0.95	0.82
2.5	0.86	0.71
3.0	0.77	0.63
3.5	0.67	0.55
4.0	0.59	0.49
5.0	0.44	0.32
6.0	0.33	0.27
7.0	0.24	—
8.0	0.18	—
10.0	0.10	—

† Based on data of Krishnaswamy (1978) for a 6701 seed normalised to 1.10 cGy cm^2 mCi^{-1} h^{-1} at a distance of 1 cm.

‡ Based on data of Ling *et al* (1983b) for a 6711 seed. Authors use a factor of 1.14 cGy h^{-1} at 1 cm from 1 mCi apparent activity to obtain dose rate variation with distance.

1.45 $R\,h^{-1}$ at 1 cm, on the transverse axis, but if the dose rate is averaged over a 2π solid angle, an anisotropy factor of 0.86 is measured. Using these factors with a roentgen-to-cGy factor of 0.90, a specific dose rate factor of 1.1 $cGy\,cm^2\,mCi^{-1}\,h^{-1}$ is obtained (Anderson *et al* 1981a). For practical dosimetry Anderson recommends treating the seeds as point sources based on the normalised transverse axis data of Krishnaswamy. For type 6711 seeds, Ling *et al* (1983b) have made further measurements which produce a different dose/distance relationship. This has been expressed by the authors as a *relative dose distribution factor*, data for both types of source being given in table 3.7.

3.6.1.3 Californium-252. Californium-252 seeds can be obtained in preloaded plastic tubes with a 1 cm spacing between active seeds. The seeds themselves have 0.25 mm of 10% iridium–platinum alloy on their walls which produces a strong angular dependence in the absorbed dose distribution for the beta- and gamma-rays. Krishnaswamy (1971) used a Monte Carlo calculation technique to establish the neutron dose delivered to tissue from a point source. The majority of the absorbed dose is due to elastic scattering of fast neutrons by the hydrogen nucleus, although a small contribution to the absorbed dose comes from the recoil of oxygen, carbon and nitrogen nuclei and from the proton released when a thermal neutron is captured by a nitrogen nucleus. Additional absorbed dose is due to the gamma-ray produced by the capture of thermal neutrons by hydrogen and the primary gamma-ray emission of californium-252. This work has been substantiated by Shapiro *et al* (1976) and Anderson (1975) who derived a specific dose rate factor for use with californium-252 after-loading seeds, type ALC-P4C, together with the specific dose rate factor for the beta- and gamma-rays produced during californium-252 decay. Due to the strong angular dependence for the beta- and gamma-rays, Anderson recommends that in dosimetry calculations the seeds be treated as line sources rather than point sources.

Other examples of non-spherical point sources used in brachytherapy include iridium-192, gold-198 and radon-222. It is customary to treat these sources as point sources and use inverse square law techniques for computational purposes, although Ling *et al* (1983a) have shown that there is a directional dependence of the radiation fluence for iridium-192 and gold-198 seeds.

3.6.2 Line sources

As discussed in §§3.3 and 3.4, the dose distribution in a medium around a line source is dependent on the construction of the source, self-absorption within the radionuclide, the filtration due to the capsule walls and the build-up effect. Line source dosimetry was first studied in the 1920s by Sievert who formulated a method of calculating the dose distribution in air around a radium tube. Interest in this topic occupied the attention of many physicists over the next 20 years until in the 1940s Quimby (1944) and Meredith (1947) published details of exposure rates around filtered radium sources of various lengths, diameters and strengths. These data are typically presented in tabular form which gives

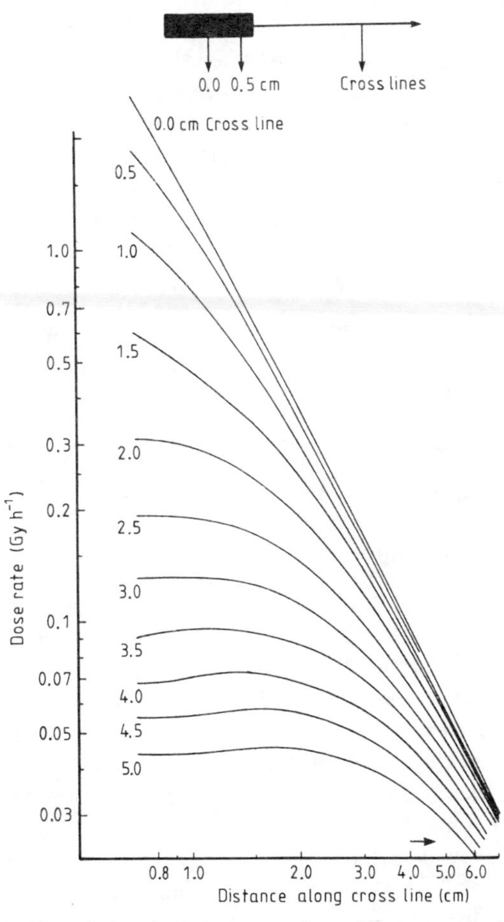

Figure 3.4 Plot of absorbed dose rate along different cross lines for a typical 20 milligram-radium-equivalent caesium-137 G-type tube. (From Klevenhagen (1973).)

the exposure rate to various points in an array around the source, or as a set of graphs which relate the exposure rate away from the axis of the source for different cross lines (figure 3.4). Whilst these workers used methods similar to that of Sievert, others have used the interval method. Interest still exists in finding methods of improving sealed-source dosimetry and Jayaraman and Agarwal (1982) have developed this interval method by considering a brachytherapy source as a cylindrical volume of radionuclide contained within a metallic shell.

3.6.2.1 Radium-226. Of all sources used in brachytherapy most consideration has been given to the distribution of dose around radium line sources. Most of the early work assumed that tissue absorption and scatter effects were small enough to be ignored and that the radioactivity was concentrated in a thin line surrounded by a cylinder of non-active material. Using the original tables of Sievert integrals, Sievert calculated the intensity of gamma-radiation, $\dot{I}_{p}$, around a radium tube. To obtain these exposure rates he divided the regions around the tube into four zones. These zones were defined such that the gamma-rays emanating from the elements were distributed as follows.

(i) Zone A: all source elements exited through the side walls.
(ii) Zone B: a proportion of the source elements exited through the side walls, whilst rays from the remaining elements exited through the end walls.
(iii) Zone C: all source elements exited through the end walls without traversing any of the radium salt.
(iv) Zone D: all source elements pass through the radium salt itself before exiting through the end walls.

For zones A and C standard tables were used appropriate to the wall thickness, whereas for zone B the exposure rate was calculated by summing Sievert integrals for the relative proportion of the active source length emitting gamma-rays through the side and end walls. Details of the equations used for each zone are discussed by Wood (1981) and Rose *et al* (1966). The early distributions tended to overestimate the filtration in the source encapsulation regions on or near the longitudinal axes and the paraxial regions. With the advancement in electronics and the use of computers, renewed attempts were made to achieve improved dose accuracy. In the late 1940s Kemp developed a dose calculator which employed an electrical method of integration and used it to calculate dose charts for 21 'approved types of radium needles and tubes' supplied by the Radiochemical Centre, Amersham (Kemp and Hall 1952). In this work Kemp introduced a 'screenage function' to allow for the fact that since gamma-radiation from radium is not homogeneous, the effective absorption coefficient of the platinum wall is a function of path length, even though the charts produced did not provide isodose information at the ends of the sources. Rose *et al* (1966), using a zone technique, provided a method of obtaining more detailed distributions which showed the reduction in dose in the region near the ends of the tube where the gamma-rays have to pass through a greater thickness of wall; the 'pinch effect'.

In an attempt to provide improved dose estimates for clinical use Young and Batho (1964) produced dose tables taking into account self-absorption of gamma-rays within the radium source and the effectiveness on the gamma-rays of the filtration due to the wall. Later in the same year Batho and Young (1964) provided a further correction to their tables to account for attenuation and scatter in tissue—$f_2(d,\theta)$. This table was later modified following the publication of Smocovitis *et al* (1967). The basis of the Batho and Young tables is the evaluation of the Sievert integral, taking into account tissue attenuation, using equation (3.16) for $f_2(d,\theta)$ and is given by

$$\dot{X}_{\mathrm{P}} = (MC/L) \int_{-L/2}^{L/2} [\exp(-\mu_{\mathrm{en}} T \sec\theta)] f_2(d,\theta)/d^2 \, \mathrm{d}L. \qquad (3.28)$$

Thus

$$\dot{X}_{\mathrm{P}} = (MC/Ly) \int_{\theta_1}^{\theta_2} \exp(-\mu_{\mathrm{en}} T \sec\theta) \, \mathrm{d}\theta$$

$$- 0.01 y^{5/4} (MC/Ly) \int_{\theta_1}^{\theta_2} \sec^{5/4}\theta \exp(-\mu_{\mathrm{en}} T \sec\theta) \, \mathrm{d}\theta \qquad (3.29)$$

where M is the radium content of the source, L is the active length and T is the effective wall thickness of the platinum. The term C is the specific gamma-ray constant for radium corrected back to zero filtration multiplied by 0.97, the roentgens-to-rads factor for soft tissue, i.e. $C = 0.97 \times 8.25 \exp(-0.05 \times 1.92) = 8.81$ cGy mg^{-1} h^{-1}. The term μ_{en} is the effective absorption coefficient and is taken to be a function of ($d \sec\theta$), as suggested by Young and Batho (1964).

By writing $x = jL$ and $y = kL$, equation (3.28) can be rewritten

$$\dot{X}_{\mathrm{P}} = [(M/L^2)(1 - L^{5/4} T(\theta_1, \theta_2))](C/k) \int_{\theta_1}^{\theta_2} \exp(-\mu_{\mathrm{en}} T \sec\theta) \, \mathrm{d}\theta$$

where

$$T(\theta_1, \theta_2) = 0.01k^{5/4} \int_{\theta_1}^{\theta_2} \sec^{5/4}\theta \exp(-\mu_{\text{en}} T \sec\theta)\, d\theta$$
$$\times \left[\int_{\theta_1}^{\theta_2} \exp(-\mu_{\text{en}} T \sec\theta)\, d\theta \right]^{-1}. \qquad (3.30)$$

By taking this approach Young and Batho were able to define an equation of the form

$$\dot{X}_{\text{P}} = (M/L^2)\{[(C/k)(U(\theta_2) - U(\theta_1))]$$
$$- L^{5/4}[0.01k^{5/4}(V(\theta_2) - V(\theta_1))/(U(\theta_2) - U(\theta_1))]\} \qquad (3.31)$$

where $U(\theta)$ are Sievert integrals as defined in equation (3.13) and

$$V(\theta_i) = \int_0^{\theta_i} \sec^{5/4}\theta \exp(-\mu_{\text{en}} T \sec\theta)\, d\theta.$$

The useful feature of this expression is the fact that, apart from the filtration factor ($\mu_{\text{en}}T$), the expressions in the square brackets are independent of the dimensions of the needle and depend only on the coordinates of P expressed in terms of the active length L of the container. In the original paper by Young and Batho (1964) a table of dose factors was given, the first square bracket in equation (3.31), from which the dose rate at P can be obtained by multiplying the dose factor by M/L^2.

To account for tissue correction, a table of tissue correction factors was derived for $T(\theta_1, \theta_2)$ from the second square bracket in equation (3.31); again the actual correction factor to be used for point P is taken from the table and multiplied by $L^{5/4}$ and the dose factor M/L^2. The corrected dose to point P is then obtained by subtracting this tissue correction factor from the uncorrected exposure rate. Since these tables are independent of the sources used, then (provided the two factors read from the tables are chosen for the actual sources used) the dose in tissue can be determined for any size source of radium.

In 1968 Shalek and Stovall recalculated tables of dose distributions in tissue around linear radium sources using the interval method with the Meisberger correction for the effective attenuation of gamma-rays in water (Shalek and Stovall 1968) and their results agreed to within about 1% with those of Young and Batho (1964). Using a more fundamental computation technique based on the recent photon interaction data, Kim (1976) showed agreement with the results of Young and Batho and Shalek and Stovall for a 1 mg radium source, but there was an apparent discrepancy for a 13.3 mg radium source filtered through a platinum tube due to photon attenuation and scatter.

3.6.2.2 Caesium-137. When caesium-137 was first introduced as a possible radium substitute the sources were encapsulated in steel. Horsler *et al* (1964) carried out measurements on these sources and reported that the caesium isodose curves were more circular than those of radium. Later, when the encapsulation was changed to 0.5 mm of 10% iridium–platinum alloy, Jones and Stacey (1965) carried out further experiments to study the differences in distribution. Klevenhagen (1973) made an experimental study of the dose distribution in water around G-type caesium-137 tubes (at the Radiochemical Centre, Amersham) and measured similar isodose distributions to those obtained by radium.

From this work isodose curves and cross-line plots were derived for an 'average caesium source'. Not only did Klevenhagen study distribution, he also compared exposure and absorbed dose rates. The activity content of caesium sources is conventionally measured in terms of milligram-radium-equivalent, i.e. the mass of radium-226 in the form of a point source screened by 0.5 mm of platinum which will give the same exposure rate in scatter-free conditions at the same distance from the source as the caesium-137 source. However, since the radium and caesium tubes are constructed differently and hence have different filtration, and since the two isotopes have different scatter–absorption relationships in water, it was found that a caesium source of 1 mg radium equivalent produces, all along the lateral axis, dose rates which are on average 9.7% higher than those for a 1 mg radium tube with 1 mm iridium–platinum alloy walls—a fact which has to be considered when caesium sources are being used to replace existing radium stock.

Using the point source summation technique, the build-up method, Krishnaswamy (1972a) produced tables of dose in rads per hour for various milligram-radium-equivalent caesium-137 sources encapsulated in stainless steel (with equation (3.26)). In this work it was determined that 2.55 mCi of caesium-137 is equivalent to 1 mg of radium when a caesium needle, encased in 0.5 mm of stainless steel, is compared to a radium source having a 0.5 mm platinum wall: this was in excellent agreement with the measured value of 2.6 obtained by Horsler *et al* (1964).

Commercially available caesium sources encapsulated in iridium–platinum alloy were studied by Breitman (1974). Using the interval method the dose rate to a point P from a linear caesium-137 source was obtained by dividing the active length into N point sources and summing the contribution from each. The absorbed dose rate, $\dot{D}_P$, is then given by

$$\dot{D}_P = \Gamma_\delta f_3(d,\theta)(A/N) \sum_{i=1}^{N} [\exp(-\mu_{en} t_i)] f_2(d_i,\theta)/d_i^2 \quad (3.32)$$

where $f_3(d,\theta)$ is the roentgen to cGy conversion factor for muscle (0.957), Γ_δ is the specific gamma-ray constant, A is the activity of the source (in mCi), μ_{en} is the linear absorption coefficient of the wall material for photons from caesium, d_i is the distance from the point P to the ith interval, t_i is the length within the filter for a ray from the ith interval and $f_2(d_i,\theta)$ is the tissue correction factor.

In Breitman's method the region around the source was divided into three zones, dependent upon the amount of wall traversed by the gamma-rays, and the value of t_i was chosen accordingly when taking into account the effective side wall thickness or end wall thickness. The correction applied to convert exposure in air to exposure in tissue, $f_2(d_i,\theta)$, was that due to Meisberger *et al*, whilst the linear absorption coefficient, μ_{en}, was taken to be 1.25 cm^{-1}, obtained from the data of Storm and Israel (1970). Correction was also made for the self-absorption of the photons with the sources. Using this approach tables were derived for a range of G-, H-, S- and A-type tubes and needles manufactured by the Radiochemical Centre, Amersham. A full range of tables can be obtained from the IAEA Publications Department (Karner Ring 11, PO Box 590, A-1011, Vienna, Austria). In determining the activity of the source the following relationship was used

$$A = a(\mu, w, AL)\Gamma_{Ra}[S/\exp(-0.05\mu_{Pt})]\Gamma_{Cs}. \quad (3.33)$$

Using $\Gamma_{Ra} = 8.25$ R cm^2 mCi^{-1} h^{-1} and $\Gamma_{Cs} = 3.32$ R cm^2 mCi^{-1} h^{-1}, the activity becomes

$$A = 2.65Sa(\mu, w, AL)$$

where S is the calibrated source strength in milligram-radium-equivalent units and a is a geometric attenuation correction factor for a wall thickness w and active length AL.

Recently, source construction techniques have changed and stainless steel is now often used to encapsulate the sources. This obviously effects the value of $f_1(d,\theta)$ used and, in the summation technique of Breitman, the value chosen for the energy absorption coefficient of the wall material. The overall effect is shown in figure 3.5 which shows a typical isodose distribution around stainless steel and iridium–platinum encapsulated sources. It will be noted that the use of stainless steel encapsulation reduces the variation of dose distribution. Note the 'pinch effect' near the source axis due to the extra attenuation by the end of the containers. This is a common feature with encapsulated line sources.

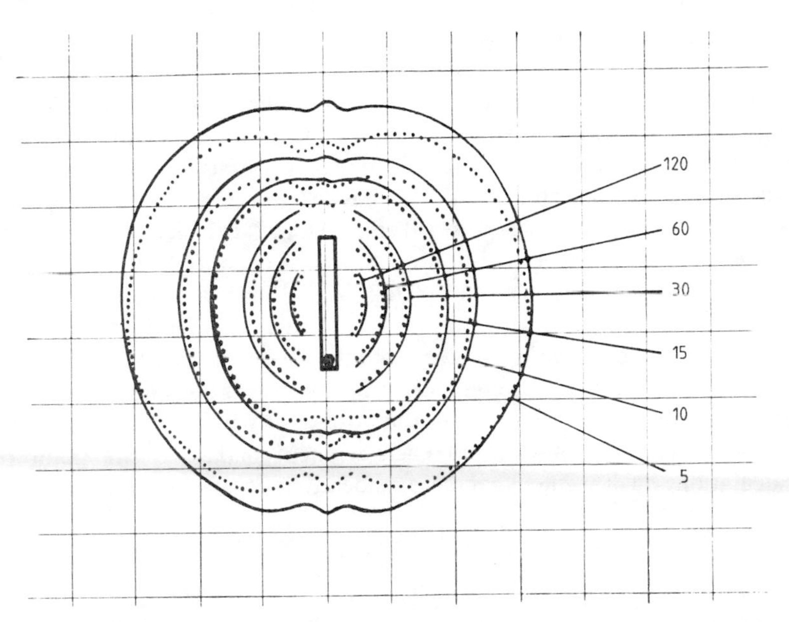

Figure 3.5 Comparison of a typical isodose distribution around a CDCS J1-type stainless steel encapsulated source (full line) and a CDCS J1-type iridium–platinum encapsulated source (broken line). Note pinch effect. Dose rates given in $cGy\,h^{-1}$. (Reproduced by kind permission of Amersham International plc.)

3.6.2.3 Iridium-192. The physical construction of iridium wires is such that the screenage function, which allows for the fact that the radiation from the source is not homogeneous, is small. In their original work, Hall *et al* (1966b) ignored this effect and assumed it to be constant because the platinum sheath is only 0.1 mm thick. The dose rate in $cGy\,h^{-1}$ in soft tissue at a point P due to a source of length L, filtered by a thickness T, is given by equation (3.33) as follows

$$\dot{X}_P = (8.3 \times 0.97 A' f/h) \int_{\theta_1}^{\theta_2} \exp(-\mu_{en} T \sec\theta)\, d\theta \qquad (3.34)$$

where $\theta_1 = \tan^{-1}[c-(a/2)]/h$, $\theta_2 = \tan^{-1}[c+(a/2)]/h$ (c is the cross line, i.e. the distance from the centre of the wire along the source axis), A' is the measured activity of the wire expressed in milligram-radium-equivalents per cm and μ_{en} is the linear absorption coefficient of the filter material. The term f is the correction factor to convert the linear activity of the wire (measured through screening) to true linear activity and T is the thickness of wire (0.15 mm) or hair-pin (0.3 mm).

Equation (3.34) was used to calculate cross-line plots for sources of different lengths. A review of these cross-line curves was made by Welsh *et al* (1983); it detailed various changes which should be applied to correct for absorbed dose in tissue. Prior to this time no correction was made for tissue absorption and this has led to errors, varying between +1.7% at 1 cm to −7.5% at 10 cm. In these calculations the value chosen for the linear absorption coefficient was $\mu = 0.43\ \text{mm}^{-1}$.

The cross-line plots derived by Hall *et al* (1966b) are for straight line sources of iridium-192 wire. In practice, the wires, when implanted, are invariably curved and thus to calculate the dose distribution around the sources a summation technique has to be applied. If the source is divided into a number of segments, then the dose to any point can be obtained by summing the contribution each segment makes to the point of calculation. Each segment can be considered to be a short line source and

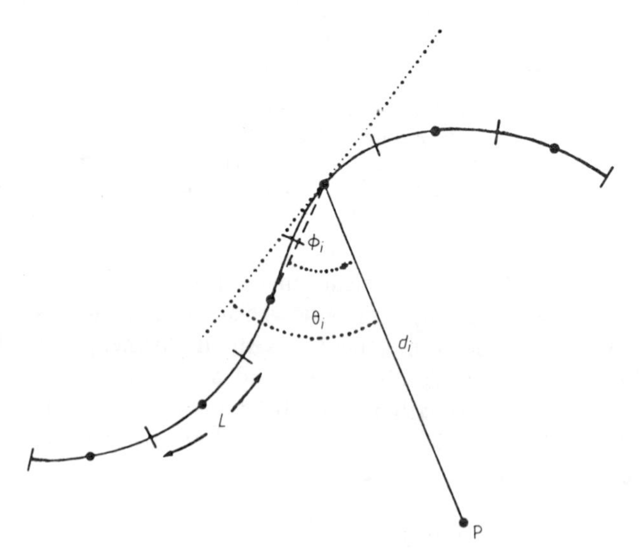

Figure 3.6 Calculation of the dose rate at a point from a curved iridium-192 wire. (After Mayles *et al* (1985).)

thus the dose is given by equation (3.8) with the added correction for $f_2(d,\theta)$ and $f_3(d,\theta)$. Since the radiation output of wire or hair-pins is typically specified in terms of A in mR h^{-1} mm^{-1} at 1 m, Mayles *et al* (1985) showed that the absorbed dose rate in water, $\dot{D}_P$ (in cGy h^{-1}), due to a curved wire (figure 3.6) is given by

$$\dot{D}_P = \sum_{i=1}^{n} AL(1000/d_i)^2 f_1(d,\theta) f_2(d,\theta) f_3(d,\theta)/f_4(d,\theta)$$

where L is the length of the segment of wire, t is the diameter of the wire in mm and θ is the angle between the line joining the point P to the mid-point of the wire segment and the tangent to the wire at the same point. This angle cannot be measured exactly but to a first approximation is equal to ϕ_i, which is the angle the line P makes with the straight line joining the mid-points of adjacent segments.

The values chosen by the authors for the various factors were as follows.

(i) The factor to account for filtration in the source in the vicinity of the source compared to the filtration under calibration conditions is given by

$$f_1(d,\theta)/f_4(s,\theta) = \exp(-\mu t/2\sin\theta_i)/\exp(-\mu t/2).$$

The factor $f_4(s,\theta)$ is discussed in §4.1.

(ii) The tissue correction factor $f_2(d,\theta)$ is that recommended by Dale (1982) (see §3.5).

(iii) The exposure to absorbed dose factor is 0.961.

3.6.2.4 Californium-252. Following the introduction of californium-252 as a brachytherapy source, various authors have published data for those distributions around various types of source (Krishnaswamy 1971, 1972b, Shapiro *et al* 1976). Most of these authors have used Monte Carlo techniques for these data by summing the dose to a point for a series of point sources. The data are either presented in tabulated form or as an isodose distribution around a tube.

3.6.3 Source trains

In many after-loading systems a linear source distribution is achieved by using a series of active and inactive beads in a source train (figure 3.7). In order to determine the absorbed dose distribution from a source train, the contribution each seed makes to any point in the vicinity of the train is obtained by using equation (3.19), with the appropriate factors to account for attenuation and scattering of the photons. The values chosen for these factors are dependent upon the path length of the ray through the capsule wall, the spiral spring and the tissue itself. Since all of these distances are dependent on the position of the source relative to the point at which the dose is being calculated, the factors for $f_2(d, \theta)$ and $f_3(d, \theta)$ have to be determined for each calculation. In addition, consideration must be given to the value used for activity in the calculation, since many algorithms require content activity rather than equivalent activity, the usual method used by manufacturers for specifying source strength. Typical dose distributions around the source trains are shown in figure 3.7, these being obtained using the method described by Cassell (1983).

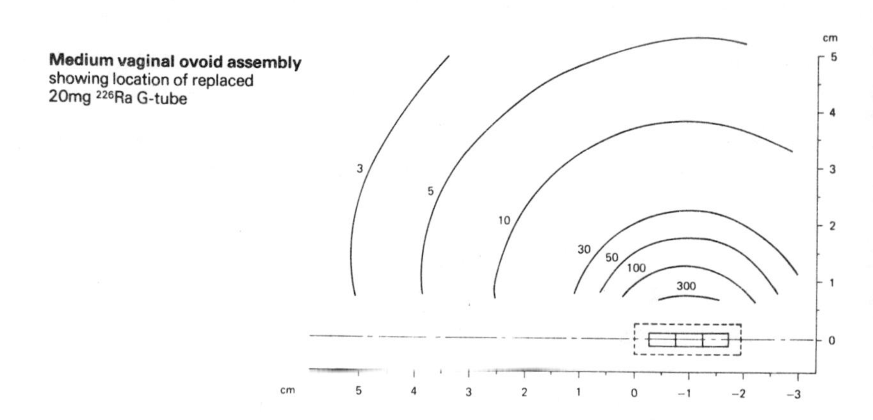

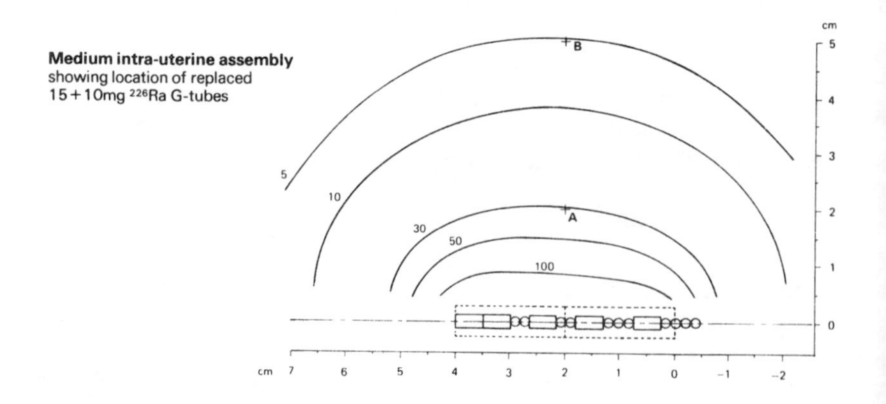

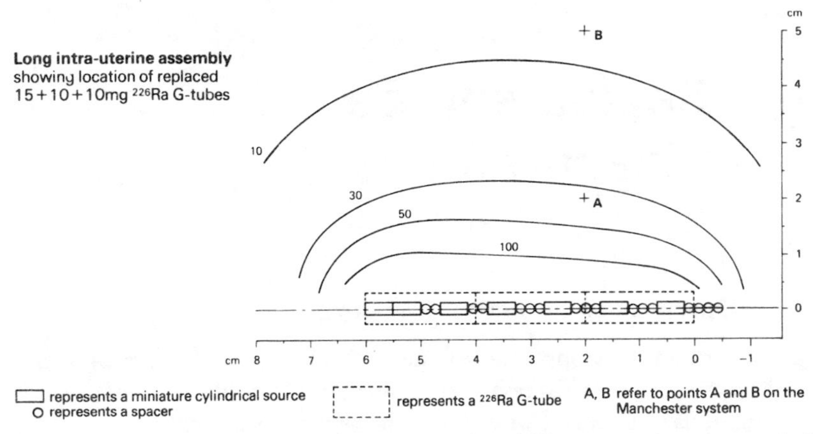

Figure 3.7 Typical isodose distribution for caesium-137 source trains composed of active and inactive pellets. (Reproduced by kind permission of Amersham International plc.)

In another study on the use of point sources to simulate line sources, Dutreix and Wambersie (1968) showed that for a series of n sources, each of activity a_0, separated by a spacing S, a linear source of length L given by $L = nS$ and of linear activity $a_1 = a_0 n/L$ was simulated. The dose distribution was found to vary little, provided the separation between sources did not exceed 2 cm; however, a dose calculation is necessary for each separation.

4 Sealed-source Measurement

4.1 Specification of Source Strength

To make any meaningful assessment of the clinical results of sealed-source therapy it is necessary to achieve uniformity in absorbed dose calculations, agreement on the way in which the strength of an encapsulated brachytherapy source is specified and traceability of source calibration to a national standard. Historically, source strength was defined in terms of the mass of radium encapsulated, together with the thickness of wall filtration. When used in conjunction with the specific gamma-ray constant, or its successors, it enabled exposure rate calculations to be made. Since the specific gamma-ray constant for radium-226 is defined in terms of 0.5 mm platinum filtration, a correction factor has to be made to account for the difference in the thickness of the wall filtration. Without this correction errors in dose of the order of 7–17% occur, depending on wall thickness, if the source strength is given in terms of the number of milligrams of radium alone.

The advent of radium substitutes has lead to much consideration of the optimum way of specifying source strength in order that the clinical experience gained with the use of radium should not be lost when other radionuclides are used in brachytherapy. Methods adopted have included specifying source strength in terms of the following:

(i) the activity of the source (content activity);
(ii) the apparent or equivalent activity;
(iii) the equivalent mass of radium, usually in milligrams; and
(iv) the exposure or air kerma rate at a specified distance from the source, typically expressed in mR h^{-1} or $\mu\text{Gy h}^{-1}$ at 1 m.

The determination of exposure rate from activity and the exposure rate constant gives rise to various problems. The activity of a source, in disintegrations per second, is not easy to measure since the isotope is encapsulated. One possible direct method employs coincidence counting techniques. Activity is, however, more usually determined from exposure measurements, but for this it is necessary to accurately know the value of the exposure or air kerma rate constant. A study of published data reveals that differences of up to 15–20% have been reported by different authors for such nuclides as iridium-192 and tantalum-182. This is perhaps not surprising in view of the complex decay schemes of these nuclides which form the basis of the determination of exposure rate constants. Additionally, the recommended use of the exposure rate constant instead of specific gamma-ray constant also causes differences, since the latter only considers the exposure rate due to gamma-rays from the source and does not include photons of characteristic x-rays and internal bremsstrahlung. Thus, specifying sources solely in terms of activity and exposure rate constant can be a source of error in brachytherapy dosimetry (Jayaraman *et al* 1983, BCRU 1984).

To overcome these problems the term *equivalent* or *apparent* activity was proposed by the IAEA (1967) and the ICRU (1970). This defines the activity of a point source of the *same nuclide* which will give the same exposure or air kerma rate in air at the same distance from the centre of the source, this distance being large enough for the actual source to be considered as a point source. Apparent activity is *not* the same as the *content activity* which is a measure of the actual source activity within the encapsulating material. An extension of this approach is to define source strength in terms of milligram radium equivalent (mg Ra eq), as mentioned before. This is defined as that activity of nuclide which delivers the same exposure or air kerma rate at the same distance from the source as a point source of radium filtered by 0.5 mm of platinum. Again this method of source strength specification presents problems since the source geometry in the two situations could be different. This highlights the differences in source specification—even though the mg Ra eq of a 1 mm and that of a 2 mm gynaecological radium tube can be made to be the same, the actual number of milligrams of radium contained in the tube—content activity—will be different. It must also be emphasised that errors will also occur if the mg Ra eq of a nuclide is determined by taking the ratio of the exposure rate constants or specific gamma-ray constants of radium and the substitute without considering the differences in filtration.

In an attempt to resolve this problem, the NCRP (1974) proposed that source strength should be specified in terms of the exposure rate at a distance of 1 m from the encapsulated source, perpendicular to the long axis of the source at its centre. This approach is now one of the recommended methods of specifying the strength of a brachytherapy source. Recently the BCRU (1984) recommended that the output of a cylindrical gamma-ray brachytherapy source should be specified in terms of the air kerma rate at a point in free space at a distance of 1 m from the source on the radial plane of symmetry, i.e. the plane bisecting the active length and the cylindrical axis of the source. In the case of a wire source the output should be specified for a 1 cm length. The recommended units for air kerma rate are micrograys/hour ($\mu Gy\,h^{-1}$). In order to convert the output of a source expressed in terms of the air kerma rate to its strength in mg Ra eq, the air kerma rate must be divided by 7.20, this conversion factor is based on an exposure rate factor of 8.25 $R\,h^{-1}\,mg^{-1}$ at 1 cm for radium-226 and assumes that an air kerma rate of 1 Gy is equivalent to an exposure of 114.5 R (see §1.2.4).

Various authors (Dutreix and Wambersie 1975, Dutreix *et al* 1982, Attix 1980, Loevinger 1981) have described ways in which the absorbed dose in a medium can be determined from such an exposure rate measurement. Since the exposure rate, $\dot{X}_s$, is measured at a large distance, the factor accounting for photon attenuation in the capsule, $f_4(s,\theta)$, at the calibration point will not be the same as the factor $f_1(d,\theta)$ (in equation (3.19)) used to calculate the wall attenuation in the vicinity of the source. In addition, a factor is necessary to correct for possible attenuation of the photons in air and photon scattering to the measuring point due to the geometry of the calibration set-up; this is $f_5(s,\theta)$. Dutreix *et al* (1982) indicated that for iridium-192 measured in air at 1 m this factor is of the order of 1.01. Using an exposure or air kerma rate measurement, the product of activity and exposure rate or air kerma rate constant is given by

$$A\Gamma_\delta f_4(s,\theta) = \dot{X}_s s^2 f_5(s,\theta) \qquad (4.1)$$

where A is the content activity, $\dot{X}_s$ is the measured exposure rate at a distance s (sufficiently large that the source can be considered to be a point source), $f_4(s,\theta)$ is the factor which accounts for the attenuation in the source capsule and $f_5(s,\theta)$ accounts for the attenuation and build-up in air.

It is also possible to define apparent or equivalent activity, A_a, using equation (4.1) such that

$$A_a\Gamma_\delta = \dot{X}_s s^2 \qquad (4.2)$$

where $A_a = Af_4(s,\theta)/f_5(s,\theta)$.

By substituting for $A\Gamma_\delta$ from equation (4.1) into equation (3.19), it is possible to define the absorbed dose in tissue in the form

$$d^2\dot{D}_P(d,\theta) = \dot{X}_s s^2 f_1(d,\theta) f_2(d,\theta) f_3(d,\theta) f_5(s,\theta) / f_4(s,\theta). \quad (4.3)$$

From this equation we see that a knowledge of the activity and the exposure rate constant or air kerma rate constant are not necessary to calculate the dose rate. The approach also has the advantage of achieving direct traceability to exposure standards maintained by a National Standard Laboratory using an exposure- or air-kerma-calibrated ionisation chamber.

For a source strength defined in terms of mg Ra eq, the exposure rate $\dot{X}_s$ due to a radionuclide source is compared to the exposure rate $\dot{X}_{Ra}$ of a mass M_{Ra} of radium. Thus the mg Ra eq, M_{eq}, of a radionuclide is defined by

$$M_{eq} = (\dot{X}_s / \dot{X}_{Ra}) M_{Ra}.$$

This relationship is only valid if the instrument reading is independent of source geometry and energy.

When sources are supplied by a manufacturer they can be specified in terms of the exposure/air kerma rate at a defined distance of say 25 cm or 1 m or as equivalent activity, which when quoted has been derived by the manufacturer from an exposure/air kerma rate measurement. The value of the exposure/air kerma rate constant used to determine this equivalent activity may not, however, be the value detailed in many recognised publications. It is therefore essential that the user knows the value used to derive equivalent activity in order that the true exposure rate can be calculated. Manufacturers also use the content activity to specify source strength. To avoid errors in dose calculation it is essential that the user knows how the source strength has been specified. To illustrate this problem, we consider the requirement to determine the content activity, for computational purposes, of a spherical source of caesium-137 supplied for use with an after-loading system. The manufacturer's test certificate specifies the source strength by stating that the air kerma/exposure rate at 1 m is equal to 43.7 μGy h^{-1} (5.0 mR h^{-1}) and that the source has an equivalent activity of 560 MBq (15.15 mCi); the manufacturer's recommended air kerma/exposure rate constants being 0.0779 μGy h^{-1} MBq^{-1} m^2 (0.33 mR h^{-1} mCi^{-1} m^2).

Example 4.1

(i) To determine content activity derived from air kerma/exposure rate measurement.

Before calculating content activity from equation (4.1) it is necessary to derive the factor $f_4(s,\theta)$; we assume $f_5(s,\theta)$ to be unity. From a knowledge of the measurement technique used by the manufacturer to determine source strength, the factor $f_4(s,\theta)$ is given by

$$f_4(s,\theta) = \exp[-(\mu(\mathrm{s})d(\mathrm{s}) + \mu(\mathrm{ss})d(\mathrm{c}) + \mu(\mathrm{ss})d(\mathrm{h}))].$$

For this type of source (figure 2.4) d(s) is the radius of the caesium-137 bead (0.055 cm), d(c) is the thickness of the stainless steel capsule wall (0.035 cm), d(h) is the thickness of the stainless steel calibration holder (0.035 cm), μ(s) is the linear attenuation coefficient for the caesium-137 glass bead (0.085 cm^{-1}) and μ(ss) is the linear attenuation coefficient of stainless steel (0.34 cm^{-1}). Therefore

$$\begin{aligned} f_4(s,\theta) &= \exp\{-[0.085 \times 0.055 + 0.34 \times (0.035 + 0.035)]\} \\ &= 0.972. \end{aligned}$$

Using equation (4.1), the content activity is given by

$$A \times 0.0773 \times 0.972 = 43.7 = 582 \text{ MBq}$$

$$\text{or } A \times 0.3275 \times 0.972 = 5.0 = 15.7 \text{ mCi}.$$

The values of the air kerma/exposure rate constant are those in table 3.1.

(ii) To determine content activity derived from equivalent activity.

Using equation (4.2), the air kerma/exposure rate at 1 m is determined by using the manufacturer's recommended values for the air kerma/exposure rate constants. Thus, for source strength in equivalent activity we have

$$560 \text{ MBq} \times 0.0779 = 43.7\ \mu\mathrm{Gy\,h^{-1}} \text{ or } 15.15 \text{ mCi} \times 0.33 = 5.0 \text{ mR h}^{-1}.$$

These derived figures for activity are then used as in (i) above.

If the equivalent activity is used in dosimetry calculations instead of the content activity then an error of 3.8% is made and a further 1% error occurs if the air kerma/exposure rate constants recommended by the manufacturers are not used.

4.2 Dosimeters for 'Absolute' Dose Measurement

4.2.1 Gamma-ray dose

Direct traceability to a National Standard Laboratory is required by the NCRP recommendations for determining source strength in terms of the exposure rate at a distance of 1 m from the source. Such a measurement is to be made under conditions of open geometry using an ionisation chamber which has a calibration factor independent of energy and whose wall thickness is such that electronic equilibrium is achieved. The procedure adopted by the National Bureau of Standards, Washington, is to calibrate, under the recommended open geometry conditions, one of the spherical standard graphite cavity chambers designed by Wyckoff and colleagues (Loevinger 1981) with representative sources of cobalt-60 and caesium-137. These calibrated sources then act as working standards against which other sources are calibrated using a 2.5 litre spherical aluminium ionisation chamber. In this method the standard source and the source to be calibrated are alternately placed in a plastic trough at the same height as the centre of the ion chamber and at a distance of approximately 1 m from it. Calibration of radium sources is essentially the same except that the working standards have been calibrated in terms of mass of radium.

Unfortunately, for routine clinical use this method of calibration is difficult and time-consuming, thus re-entrant ionisation chambers, initially designed for assaying liquid source activity, are used. Such well-type re-entrant ionisation chambers (figure 4.1) consist of an inner and outer cylinder with base plates which act as the HT electrode and an intermediate cylinder, insulated from the base plate, which acts as the collecting electrode. The sources are placed in Perspex holders which allow them to be located towards the centre of the cavity. The gas within the ion chamber is dependent on chamber design; it can either be air at atmospheric pressure (as in the case of the type 1383A chamber), argon at 10 atmospheres (for the NPL/Vinten Model 671 chamber) or argon at 12 atmospheres (in the case of the Capintec CRC-10 isotope calibrator).

For the calibration of gamma-ray sources it is essential to remove all beta-particles. This is achieved by using brass or Perspex liners. In order to ensure that the ions produced within the chamber do not recombine, the polarising voltage between the electrodes is of the order of 200–600 V. It is also essential to remember that the chamber reading is likely to be increased in use due to scattered radiation from the lead which is placed around the outer cylinder to reduce the effects of extraneous radiation. Dale *et al* (1961) studied the characteristics of the 1383A re-entrant chamber and noted, among other things, that the increased ionisation due to this effect could be as much as 3%, depending on the thickness of lead used and the radionuclide being studied. It has also been found that the response of these chambers is a function of the energy of the gamma-rays, the position of the source in the cavity and the dimensions of the source. Depending on chamber design, it may be necessary to apply an atmospheric correction factor and account for the fact that the chambers may not have a linear response in relation to source activity.

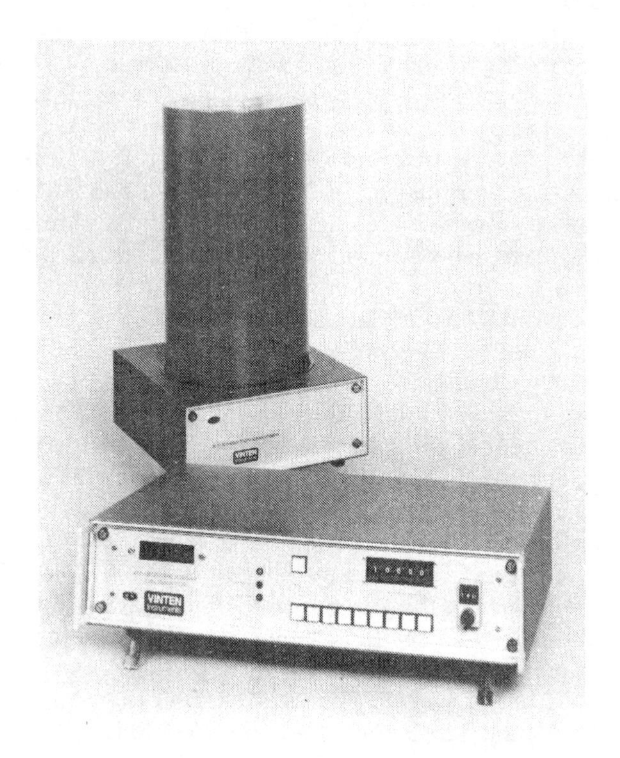

Figure 4.1 Re-entrant ionisation chamber. (Reproduced by kind permission of Vinten Instruments plc.)

By considering an isotropic point source placed at the centre of an idealised well chamber geometry, Boyer *et al* (1981) were able to show how the above factors affect the current measured. A typical curve showing the variation in ionisation current in pA MBq^{-1} with photon energy is given in figure 4.2. To obtain the total current per MBq of radionuclide it is necessary to sum the contribution the photons of each energy make to the total. Typical values obtained with sources used in brachytherapy are given in table 4.1 for the 1383A chamber converted to 760 mm of Hg

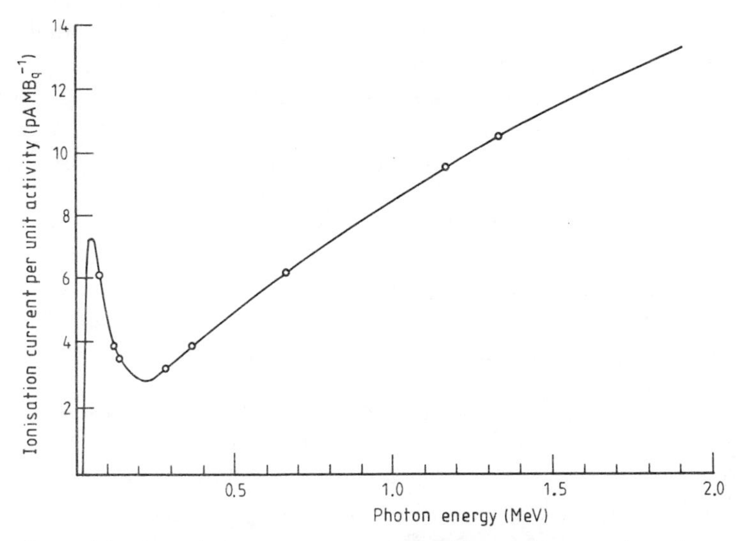

Figure 4.2 A typical current per unit activity against varying photon energy curve for a re-entrant ionisation chamber.

and 22°C and for the 671 chamber which requires no temperature and pressure corrections.

Re-entrant ionisation chambers also exhibit an output variation which is dependent upon the position of the source within the chamber. This variation is a function of the size of source and its distance from the base

Table 4.1 Ionisation current produced per unit activity.

	1383A chamber†		671 chamber‡	
Radionuclide	Without liner (pA mCi^{-1})	With liner (pA mCi^{-1})	5 ml BS ampoule (pA MBq^{-1})	10 ml P6 vial (pA MBq^{-1})
Cobalt-60	56.7	52.9	22.27	22.17
Caesium-137	15.8	13.7	5.79	5.75
Iridium-192	26.9	22.7	8.50	8.44
Gold-198	12.5	10.8	—	—
Radium-226	—	34.5	—	—

† Dale *et al* (1961), Dale and Williams (1964), Woods (1970), Woods and Lucas (1974) and Woods (private communication, 1978).

‡ Users Manual, model 271/671 Isotope Calibrator, manufactured by Vinten Ltd (Woods *et al* 1983).

of the chamber. For sealed-source dosimetry it is essential that such variations are determined prior to using the re-entrant chamber for source calibration. Many workers, including Dale *et al* (1961), Cobb and Bjarngard (1974), Berkley *et al* (1981), Kubiatowicz (1981) and Williamson *et al* (1982, 1983a and b), have described the use of various re-entrant ionisation chambers for determining the activity of a source. By using a calibrated radium source it is possible, by applying the necessary corrections to account for position in the chamber and energy response, to obtain the ma Ra eq of a source.

A further method for determining the activity is to carry out a spectrometric analysis of the source using, for example, a high-purity Ge detector (Cobb *et al* 1981, Boyer *et al* 1981, Ling *et al* 1981). In this technique the Ge detector is connected to a 4096 multichannel analyser and from the resulting energy spectrum the total number of photons from each energy level can be determined, making due allowance for the energy-dependent intrinsic efficiency of the detector. The activity of any source can then be determined from the gamma-ray spectrum of the isotope using the following equation for each selected energy peak:

$$A = C_{\mathrm{N}}[t\varepsilon fG \exp(-\mu x)]^{-1} \quad \mathrm{s}^{-1}.$$

In this equation C_{N} denotes the net counts for a particular energy peak minus background, t is the counting time in seconds, ε is the internal efficiency of the detector, f is the number of unconverted photons per disintegration, $\exp(-\mu x)$ is the attenuation in air and the detector window and G is given by $G = S/4\pi x_{\mathrm{d}}^2$, a geometry sensitivity factor for the chamber. In the above expression for G, S is the area of the detector face and x_{d} is the source/detector distance.

Using this technique it is possible to determine the activity of even low-energy sources such as iodine-125 (Eldridge and Crowther 1964, Ling *et al* 1981). For this isotope it has been accepted that the activity is arbitrarily defined in terms of millicurie compensated (mCi Comp) where 'A 1.0 mCi Comp source contains sufficient iodine-125 that, when assayed using exposure rate measurements at points along an axis of the source, the source would appear to be a naked source of 1.0 mCi.'

4.2.2 Beta-ray dose

The determination of the absolute dose rate due to a beta-ray source, such as an applicator, is usually made using an extrapolation chamber. The requirements of such a measurement are a small collecting area and a shallow air gap and, providing these are achieved, accuracy of better than $\pm 2\%$ can be obtained (Loevinger 1953). Various authors, including Haybittle (1955), Loevinger and Trott (1966) and Payne and Waggener (1974a), have described such chambers, Failla (1937) making the first chamber to work on the extrapolation principle. In this technique the applicator is placed in position on the surface of the extrapolation chamber, but without pushing against the window and distorting it. Measurements are usually made of both the positive and negative ion currents flowing across the chamber for a given electrode separation using an electrometer, and an average ion current for a given electrode separation is thus determined. The separation is then varied using a micrometer screw and the resultant ion currents measured and plotted against plate separation and extrapolated to zero spacing. Typically, for small separations, a linear relationship exists between electrode separation and current, the slope of which can be used to determine the ionisation produced per cm^3. This value can then be corrected for (a) attenuation in the window, C_1, and (b) atmospheric temperature and pressure variations, C_2. The exposure rate at the surface of the applicator can be determined by:

$$\dot{X} = (\Delta I / \Delta d)(1/\text{area})C_1 C_2 C_3$$

where $\Delta I / \Delta d$ is the slope of the current against electrode separation graph (in A/mm) and C_3 is the conversion factor to convert ionisation current in A/mm to roentgen/s.

4.3 Dosimeters for Relative Dose Measurement

The determination of absorbed dose within tissue not only requires a knowledge of the source strength but also the dose distribution around the source. Such measurements are difficult to make, mainly due to the steep dose gradients within the region of interest from the source (5–20 mm) and the variation in the energy spectrum from different nuclides at different distances from the source. It is therefore desirable that dosimeters should be small in size and have the following response characteristics.

(i) They should have a linear relationship with absorbed dose.
(ii) They should be independent of dose rate.
(iii) They should be independent of energy.
(iv) They should be independent of angular distribution.
(v) They should have reliably stable calibration with time.
(vi) They should be easy to use.

To meet these various requirements, various authors have made use of differing dosimeters for relative dose measurements around sources, including ionisation chambers, solid state detectors, photographic films, thermoluminescent dosimeters and scintillation detectors.

4.3.1 Photographic films

Much of the early work on measuring the relative dose distribution around sealed sources was carried out using photographic films. Tochilin (1955) described a technique for measuring the distribution of dose from radium needles and plaques using a Kodak Eastman Type M industrial x-ray film, whilst Haybittle (1957) and Shalek *et al* (1957) used photographic techniques to determine the dose distribution around tantalum-182 hair-pins and gold-198 seeds respectively. Photographic film, whilst having good spatial resolution, has a response which is energy dependent and non-linear with absorbed dose, although the latter effect can be minimised by the choice of film.

In the interaction of photons with photographic emulsion, which consists of ionic silver halogens dispersed in gelatine, the ionising radiation causes silver grains to be formed, which on subsequent development produce an image of the radiation distribution. Since silver and bromine ions have a high atomic number, Z, then for photons interacting by the photoelectric effect, the degree of blackening per incident photon flux will be much greater than for photons interacting by the Compton effect, the relative sensitivity at 40 keV being of the order of 25 times greater than for cobalt-60 gamma-rays. To determine the degree of blackening, the film has to be read using a densitometer which evaluates the optical density, OD, defined as $\log_{10}(I_0/I)$, where I_0 and I are the incident and transmitted light intensities respectively. By calibrating the film under known energy conditions against dose, it is therefore possible to relate OD to absorbed dose, allowance being made for the background optical density—the fog level.

Determining the dose distribution can effectively be achieved either by manually determining the OD at each point in the film or by using a scanning microdensitometer (Planskoy 1983). The shape of the OD versus dose calibration curve is a function of the characteristics of the film, which are governed by such parameters as the grain diameter of the silver halide crystals, the number of grains per unit area and whether or not the film is coated with emulsion on one or both sides of the base. Development techniques also affect the shape of the calibration curve, whilst its slope is also dependent on the photon energy. Films which have been found to be useful for dosimetry purposes are 'slow films', those having a fine grain structure and a thin emulsion layer, since these have a reasonable linear dose response. Typical films chosen for these purposes are the industrial line films, Kodaline or Translite, although it is often advantageous to use films which are accepted by an automatic processor. One such film is Kodak RP/V film which is sold in ready-packed envelopes which will not admit light. This film has the linear portion of the dose against OD curve in the OD range 0–1.5, which corresponds to an absorbed dose in the range 0–80 cGy. However, Planskoy (1983) has shown that this film has a marked increase in sensitivity to photons whose energy is of the order of 35 keV.

Whenever photographic film is used for dosimetry it is imperative that a calibration curve is produced, since the shape and slope of this curve varies from batch to batch of film and also with processing conditions. Methods of film dosimetry are more fully discussed by Dudley (1966) and Herz (1969).

4.3.2 Thermoluminescent dosimeters (TLD)

These dosimeters have been developed over the past 20 years (Fowler and Attix 1966, Cameron *et al* 1968, McKinlay 1981, Horowitz 1981). In the process of thermoluminescence, the detector (usually made of lithium fluoride) absorbs energy from the radiation field, which causes a small proportion of electrons to be excited into the conduction band in which they travel until trapped in energy states in the forbidden zone, remaining there until being released by subsequent heating.

In the heating process the electrons are given sufficient energy to raise them to the conduction band, from which they then return to the bound energy levels with the emission of visible light. By integrating the amount of light emitted during the heating process, a measure of the energy deposited during irradiation—absorbed dose—can be obtained. This heating process and light output measurement are carried out in a TLD reader. The detectors themselves are commercially available in the form of crystals or powder. Crystals are much more convenient for routine use and come as pure lithium fluoride crystals in the form of cylindrical rods of dimensions $1 \times 1 \times 6\ mm^3$ or flat chips of dimensions $3.2 \times 3.2 \times 0.9\ mm^3$, although they are also available as lithium fluoride embedded in Teflon discs.

Prior to use, thermoluminescent dosimeters need to be exposed to defined doses of radiation to obtain a calibration curve of thermoluminescence versus absorbed dose. In use, it is customary to take the average read-out value of three to five detectors which have been irradiated for each calibration point, the absorbed dose at the calibration point being determined by an ionisation chamber measurement which gives traceability back to a National Standard Laboratory. Typical calibration curves show a linear dose response up to several hundred cGy, above which the supralinearity effect of these detectors becomes apparent. The shape and slope of the calibration curve is dependent upon such parameters as the previous radiation history of the crystals and the heating cycle to which they have been subjected. This means that the history of a batch of detectors is important. All detectors, therefore, have to be annealed together in order to maintain any degree of reproducibility of result.

Since lithium fluoride (LiF) has an effective atomic number (Z) of 8.14, the thermoluminescence response versus absorbed dose in water is only slightly dependent on energy, i.e. the relative sensitivity at 40 keV is only 25% greater compared with cobalt-60 photons. This makes LiF suitable for certain brachytherapy dosimetry applications.

Thermoluminescent dosimeters were used by Kartha *et al* (1966) to determine experimentally the absorption and build-up factors in water for various radionuclides. Lin and Cameron (1967) described the use of TLD crystals ($1 \times 1 \times 2\ mm^3$) to determine the radiation distributions around linear radium sources at points corresponding to those in the Quimby tables (see Chapters 5, 6 and 7). Also, TLD crystals were used by Krishnaswamy (1978) to verify dose calculations around iodine-125 seeds by making measurements at various distances along the transverse axis of an iodine-125 seed, although in this work the assumption was made that there was negligible variation in the photon spectral distribution at different distances from the source.

4.3.3 Scintillation detectors

The ability of ionising radiation to cause flashes of light or scintillations in suitable crystals has been used for over 30 years as a means of measuring the brachytherapy dose distribution. The crystal is connected either directly, or through a light guide, to a photomultiplier which converts the light flashes into electron pulses whose amplitude is proportional to the energy deposited in the scintillator by the ionising photon. The pulses are then either counted or converted into a current which is proportional to the dose rate at which energy is absorbed by the crystal. These devices are extremely sensitive to radiation; dose rates as low as $10\ \mu Gy\ h^{-1}$ can be measured using an anthracene crystal of $1\ cm^3$. Since for brachytherapy purposes dose rates are typically in the range 1 to $500\ cGy\ h^{-1}$, the volume of the detector can be made small to achieve good spatial resolution. These devices are, however, energy dependent, this being particularly so for crystals which have high atomic numbers (such as thallium-loaded sodium iodide).

Various crystals have been used for relative measurements in brachytherapy dosimetry. Some of the early work was carried out by Hine and Freidman (1950) who used a 3.2 mm diameter, 3.2 mm thick calcium tungstate crystal connected to a photomultiplier as the radiation detector when using an automatic isodose recorder to measure the dose distributions around a linear radium source in air and in water. This type of crystal has a large energy dependence, thus other plotter systems were devised which used an anthracene crystal which is much less energy dependent (Cole *et al* 1953, Fletcher *et al* 1954). Meisberger *et al* (1968), in their measurements to establish the effective attenuation in water of the gamma-rays from various radionuclides, used a cylindrical anthracene crystal 3 mm in diameter and 3 mm in length coupled to a photomultiplier by a long Lucite light guide. In such measurements care has to be taken to shield the photocathode of the photomultiplier from the radiation emanating from the source, and corrections must be applied to compensate for the dark current from the photomultiplier and any light generated in the light guide due to Cerenkov radiation.

Inorganic scintillation counters have also been used in brachytherapy measurements. Horsler *et al* (1964) compared the results they obtained using photographic film with measurements made using a scintillation dosimeter constructed of a plastic phosphor, 10 mm long and 5 mm in diameter, loaded with zinc sulphide powder. In this study the light output was assumed to be directly proportional to exposure rate, the basis of the assumption being due to the work of Belcher and Geilinger (1957). The ability of scintillation detectors to determine spectral distribution has also been widely used to establish the contamination of an isotope by another isotope (Horsler *et al* 1964) and as a means of determining the exposure rate constant for an isotope (Schulz *et al* 1980). A general review on the use of scintillation detectors has been provided by Ramm (1966).

4.3.4 Ionisation chambers

Ionisation chambers connected to a suitable current- or charge-measuring electrometer probably provide the most reliable dosimeters used in radiotherapy. Constructed, in general, of air or water equivalent walls, thimble ionisation chambers have the properties of dose rate and dose response linearity and are almost energy independent. Typical currents produced by a 1 cm^3 thimble chamber are of the order of 300 pA for an exposure rate of 1 $R\,s^{-1}$. In brachytherapy, however, we are interested in measuring exposure rates as low as 3–5 $R\,h^{-1}$, i.e. of the order of 0.01 $R\,s^{-1}$. Since current measurement below 1 pA becomes increasingly difficult, this places certain restrictions on the minimum size of ionisation chamber which can be used. Indeed, the use of ionisation chambers for measuring exposure rates in the areas of steep dose gradient around a brachytherapy source is difficult due to the size of chamber necessary to produce adequate current for accurate measurement (Klevenhagen 1973).

Whilst thimble ionisation chambers have been used for brachytherapy dose measurements (Meredith *et al* 1966, Smocovitis *et al* 1967), the experimenters have needed to apply a correction factor to account for the variation in exposure across the chamber. In addition to this correction, it must be remembered that in many applications thimble chambers will require a build-up cap, since the energy of the gamma-rays from the nuclides is such that extra wall material is required compared to a normal thimble chamber in order to achieve electronic equilibrium.

In the measurement of dose distribution close to a californium-252 interstitial implant source, it is necessary to measure both the neutron and gamma-ray dose components in the vicinity of the source. Bednorel *et al* (1976) have used an iso-octane-filled liquid dielectric ionisation chamber for this purpose. Their research has shown that such a detector, with the appropriate instrumentation, provides good discrimination between the radiation components of such a mixed field of radiation and also possesses excellent sensitivity without producing unnecessary perturbation of the radiation field. More general surveys of the theoretical and practical aspects of the ionisation chambers have been given by Boag (1966), whilst Planskoy (1983) provides a summary of the main problems of dose measurement using ionisation chambers.

4.3.5 Solid state detectors

To overcome some of the size limitations of ionisation chambers, solid state detectors have been used by many authors as a method of measuring relative dose distributions around brachytherapy sources (Klevenhagen 1973, Diffey and Klevenhagen 1975). A solid state detector may be considered as an ionisation chamber in which the gas has been replaced by a solid. In the case of semiconductor junction (p–n) detectors, the collecting volume can be much smaller, 0.1 cm^3 or less. Since solid silicon has a density 1850 times that of air, and also the average energy required to produce an ion pair is only 1/10 of that for air, the ionisation yield per unit volume is of the order of $2\text{–}3 \times 10^4$ times greater than in an air-filled-cavity ionisation chamber.

The silicon p–n junction surface barrier detector used by Klevenhagen (1973), in conjunction with a low-input impedance amplifier, had a sensitivity of approximately 30 pA per $\mathrm{R\,h^{-1}}$. However, the main disadvantages of silicon detectors compared with air ionisation chambers are as follows.

(i) They are energy dependent, i.e. they are six to seven times more sensitive at 40 keV compared to cobalt-60 photons.
(ii) They exhibit an angular dose response.
(iii) They show a temperature dependence which is a function of the individual detector.

The latter effect can be compensated for by adjusting appropriate electronic circuit parameters (Klevenhagen 1976, 1978). To minimise the energy dependence effects of these detectors, which have a factor of two greater response for photons of energy between 20 and 200 keV, Klevenhagen (1973) only used the p–n detector to measure the relative dose variation on a series of circles centred on the middle part of the source.

Other solid state detectors suitable for dosimetry measurements are p–n junction photovoltaic detectors (Whelpton and Watson 1963). These measure the open current potential difference generated by the separation of the ions in the junction. Photoconductive dosimeters are based on the use of a cadmium sulphide crystal (Bryant 1966) but tend to have the disadvantage of variations in response with temperature and energy, as well as being susceptible to radiation damage. However, the ability of cadmium sulphide to amplify the ionisation current by three or four orders due to the trapping of electrons (leaving impaired 'holes' present in the crystal) does mean that the detector size can be small. Unfortunately, these detectors have now been found to be unsatisfactory for use in brachytherapy work. Another form of detector has recently been evaluated based on the use of diamond (Planskoy 1980). These detectors are still at the experimental stage but hopefully will provide a further method of measuring the low dose rates found in brachytherapy dosimetry. A review of solid state detectors is given by Fowler (1963, 1966).

4.4 Tube Source Distributions

Using one or other of the various measurement techniques described above, various authors have determined experimentally the dose distribution around various brachytherapy sources. When using photographic film, attention has be paid to the energy response of the film for the photons being measured. It is also essential that any possible electron contamination within the radiation field must be prevented from reaching the film. These errors can be avoided by exposing the film in a unit density material such as Temex rubber, 'Mix D' or Perspex. By taking these precautions, Haybittle (1957) and Shalek *et al* (1957) determined the dose distribution for tantalum-182 wire hair-pins and gold-198 grains, whilst Horsler *et al* (1964) and Jones and Stacey (1965) measured dose distributions around caesium-137 intracavitary and interstitial sources using an Ilford line film. Similarly, Buchan and Griffin (1969) were able to demonstrate the pinch effect around a radium tube (due to the extra photon absorption in the wall material at the ends of the radium tube) by interpolating the distribution obtained on films taken normal to the source axis and at different distances from the end of the tube.

Dose distributions measured using silicon surface barrier p–n detectors have been produced by Klevenhagen (1973). In this work 30 caesium G-type tubes (manufactured by Amersham International plc) were investigated. All tubes showed symmetrical dose distributions along the longitudinal axis but asymmetry about the lateral axis, the latter varying from source to source. To achieve these spatial distribution measurements the authors set the detector at various distances from 1 to 8 cm from the source within a water phantom and rotated the source, taking readings of the semiconductor current at 37 selected source orientations. The readings taken at a given angle were then related to the reading on the lateral axis and the relative dose rate distribution from a single caesium source on circles of 1.5 and 3 cm radius were produced. Using these results, Klevenhagen was able to compare the dose distribution for a radium tube having a 1 mm iridium–platinum alloy wall thickness with a caesium tube of 0.5 mm iridium–platinum alloy wall, assuming equality of dose rates on the central axis. In the majority of these measurements, ionisation chambers have been used to establish the dose rate at different distances from the source along the lateral axis, and these in turn have been used to normalise the relative dose rate measurements made around the tubes with more sensitive and/or smaller detectors.

4.5 Point Source Distributions

As with line sources, the dose distribution around a point source is a function of source construction. For spherical sources, with uniform wall thickness, the dose distribution is spherical. However, certain source constructions result in varying degrees of anisotropy. Diffey and Klevenhagen (1975) measured the dose distribution around a CDCK-type caesium-137 source (from Amersham International, see figure 2.4) using the approach adopted by Klevenhagen (1973) for measuring caesium-137 line sources. A typical isodose for such a source does not exhibit the pinch effect at points lying near the longitudinal axis, probably due to the hemispherical ends of the sources.

Experiment has shown that these sources cannot be assumed to have a spherical distribution, the ratio of the dose at an angle to the dose on the lateral axis being lower than that for a spherical distribution by up to 8%. A greater anisotropic effect is seen with iodine-125 seeds. Ling *et al* (1979, 1983) measured the radiation distribution around an iodine-125 seed using either a silicon–lithium-drifted detector or intrinsic Ge detector positioned at a distance of greater than 50 cm from the source. The approach permitted good circular resolution and, by correcting as necessary for the spectral response of the detector, the authors derived an effective photon angular distribution by weighting the relative calibration of each photon energy to a given angular position, which included a factor to account for the differential attenuation of the x-rays and gamma-rays by the capsule. This effective photon angular distribution has then been used to calculate the dose distribution around an iodine-125 seed. The relative dose distribution of iodine-125 has also been measured by Krishnaswamy (1978), using a Kodak type M photographic film. In this work the film was used to measure the depth dose along the longitudinal axis by sandwiching it between two sheets of Perspex with the seed source being kept at the centre in a small cavity adjacent to the film. Relative dose distributions were obtained for lines perpendicular to the source axis and at different distances from the centre. The usual dose distribution was then determined by extrapolation. A similar angular distributional effect has been reported for iridium-192 and gold-198 seeds (Ling *et al* 1983a) using a NaI crystal connected to a multichannel analyser. The results of this experiment indicate that along the seed axis the radiation fluence is of the order of 80% of that perpendicular to the seed axis.

4.6 Quality Control of Clinical Sources

To ensure accurate brachytherapy a knowledge of source activity at the time of operation is essential. When sources are supplied by the manufacturers they are usually accompanied by a certificate of measurement detailing the activity of the particular source or, where appropriate, the linear activity of the source. As mentioned above, the user must know whether the activity quoted is the content or equivalent activity. However, it is advisable that source activity should be verified locally prior to use. In the case of intracavitary tubes and interstitial needles, each source is provided with a means of identification, often in the form of a serial number engraved on the source together with the appropriate catalogue number. Tube sources are also engraved on the flat of the eyelet end with a source activity identification code.

As indicated in §4.2 traceability of activity to a Standard Laboratory can be achieved by using a calibrated isotope calibrator. Such instruments provide a simple method of ensuring that the manufacturers' stated activity agrees with local determination of activity, and thus any possible error in dispatch from the supplier can be eliminated. With long half-life radionuclide sources such a check can be carried out when the sources are delivered and then at regular intervals subsequently. With short half-life nuclides it is necessary to carry out this procedure with each consignment of isotope received. Thus a calibration curve is needed for each isotope used in the calibrator, but it must be remembered that account must be taken of the source construction in deriving this curve. In practice, it is usual to achieve this curve using a series of sources of high activity of identical construction to the sources routinely used; that is, having the same active length and wall filtration. With re-entrant ionisation chambers it must also be remembered that the position of the source within the chamber affects the output, thus this has to be reproducible each time a source calibration is made. This can be achieved by constructing a suitable plastic jig to insert in defined geometry conditions within the chamber.

The linear activity of a source can be checked using film techniques. This approach ensures that no obvious variations exist in linear activity due to the movement of the source within the capsule. Again this check should be carried out when new sources are received and then at regular intervals, even though the uniformity of distributions of the radioactive material within each source is subject to strict quality control checks during manufacture, to ensure that the activity along the length (measured using a 1 mm detector window) does not vary by more more than ±5%.

This method of quality assurance is not appropriate for use with iridium-192 wire.

Often, lengths of wire are used in several applications and any bending of the wire during insertion or on removal can cause small areas of low linear activity. To provide more detailed information on activity, a linear activometer or electrometer has been developed (Pierquin *et al* 1978a, Dutreix *et al* 1982). This consists of a radiation detector in the form of an ionisation chamber or Geiger–Müller counter placed at the end of a lead collimator which allows a length of between 4 and 7 mm of the active wire to be measured at any one time. In this procedure the wire is passed beneath the collimator and the variation in the output of the detector monitored for any variation in linear activity.

Whilst the above measurements are necessary to obtain an accurate knowledge of sources prior to clinical use, further checks are needed to ensure that sources do not become damaged and thus dangerous. Routine tests therefore have to be carried out on the sources to look for signs of damage; these checks will be considered in Chapter 9.

5 External Applicators and Moulds

5.1 Introduction

Within 20 years of the discovery of radium, it was being used on external applicators for the treatment of superficial lesions. At the time there was little appreciation of how sources should be distributed to achieve a uniform dose within the lesion; indeed, Wickham and Degrais (1910) recommended placing the sources in a uniform manner over the surface applicator to achieve a uniformity of dose within the tissue below the applicator—a recommendation which was subsequently shown to be incorrect.

Attempts to improve radium dosimetry began in the 1920s when workers determined the dose distribution around various source arrangements either by calculation or measurement. It was soon realised, however, that a general solution was required to the problem of how much radium should be used and how it should be arranged in order to achieve a defined dose on the surface of the lesion. One of the earliest attempts to find a solution was carried out by Quimby (1922). In this work various shapes and sizes of applicators were studied and the dose to the centre of the skin area below the applicator was determined by considering each source in the applicator to be a series of point sources and summing the contribution to that point for each of the point sources. Using this same method, Quimby developed a dosage system which determined the relative amounts of radiation delivered to the central point for a constant amount of radium distributed uniformly over circular, square and rectangular applicators of various areas. Placing radium uniformly over the surface of an applicator, however, gives poor uniformity of dose over the area beneath the applicator. The development of the mathematical approach to radium dosimetry in the 1930s saw consideration being given to the uniformity of dose over the treatment area, as well as the dose delivered. Paterson and Parker (1934, 1938), working at the Christie Hospital, Manchester, adopted this approach and derived a series of tables which achieved a dose uniformity of ±10% over the treated surface, provided certain distribution rules were carefully observed. These rules and tables form part of the Paterson–Parker or Manchester radium dosage system, which was subsequently published in book form by Meredith in 1947 (Meredith 1947).

Surface applicators were used extensively for many years in the treatment of superficial lesions of, for example, the ear, lip, floor of mouth, forehead, penis and limbs. However, many of the lesions formerly treated by moulds are now treated using other modalities of radiation, such as electron beam therapy. For clinical use the mould is individually made for each site to be treated such that:

(i) the distance from the source to the skin is accurately maintained over the surface area;
(ii) the applicator can easily and accurately be fixed to the patient; and
(iii) the sources are securely fixed so that the loss of a source from the mould is extremely unlikely.

Materials typically used for applicator construction include plaster, wax, elastoplast, felt, Perspex, dental compounds and thermoplastics. A thorough account of the construction of applicators and their clinical uses can be found in Paterson's *The Treatment of Malignant Disease by Radiotherapy* (Paterson 1963) and *The Radiotherapy of Malignant Disease* by Easson and Pointon (1985).

5.2 The Paterson–Parker or Manchester Dosage System for Applicators

5.2.1 Dosage tables

The original dosage tables derived by Paterson and Parker give the amount of radium in milligram–hours (mg h) required to give a dose of 1000 R to the treatment area which is at a distance h from the surface of the applicator, whilst at the same time achieving a dose uniformity over that area of ±10%. Until this time the roentgen had only been considered to be applicable to the measurement of the exposure due to x-rays, but in this work the concept of using the roentgen as a unit of dosage was extended to gamma-rays. In the derivation of the tables two basic assumptions were made, namely (a) the specific gamma-ray constant for radium, Γ_δ, was 8.4 $\mathrm{R\,h^{-1}\,mg^{-1}}$ at 1 cm, and (b) the exposure rate from a source is a function of the inverse square law alone and no allowance need be made for the attenuation and scattering of the photons within the mould or tissue.

It is now recognised that these basic assumptions are only a first approximation and that the tables need correcting to enable the absorbed dose within tissue to be calculated more accurately. To derive a correction factor, account has to be taken of the recent determination of the exposure rate constant for radium, differences in exposure due to photon attenuation and scatter, filtration of the photons in the walls of the containers, the variation in attenuation coefficient with wall thickness of platinum, as well as the factor relating exposure and absorbed dose. Various authors have suggested corrections ranging from 1.064 (Johns and Cunningham 1983) to 1.11 (Stovall and Shalek 1968), whilst Porter (1970) concluded that the average value of the Paterson–Parker '*R*' in brachytherapy was equivalent to '0.92 rad' and recommended that the tables should be used with this factor. Recently, workers at the Christie Hospital have recommended a factor to correct the original tables to account for these differences (Gibbs and Massey 1980). This includes

(i) a factor of +1.5% to account for filtration and attenuation and scattering differences,
(ii) a factor to incorporate the current exposure rate constant for radium of 8.25 $\mathrm{R\,cm^2\,h^{-1}\,mg^{-1}}$ and
(iii) an exposure to absorbed dose factor of 0.957.

These corrections yield an overall factor of 1.08 by combining as (8.4/8.25)(1/0.957)1.015 = 1.08. By multiplying the earlier dosage tables of Paterson and Parker by this factor the number of mg h required to deliver a dose of 10 Gy is obtained. The recently derived dosage table for surface applicators is given in table 5.1, the data being presented in tabular form relating the number of milligram–hours per 10 Gy for various areas and treating distances, assuming a filtration of 0.5 mm platinum. To account for any changes in wall thickness, a correction factor of ±2% per 0.1 mm extra (or less) platinum must be applied. To follow the recommendations of the BCRU (1984)—that source strength should be specified in terms of air kerma at 1 m—table 5.1 has been recast in the form of air kerma at 1 m per unit of system absorbed dose in units of $\mu\mathrm{Gy\,Gy^{-1}}$, the conversion factor being 0.72 (Massey *et al* 1985). This value accounts for the difference in specifying 'dose' in terms of 1 Gy rather than 10 Gy and the fact that air kerma rate for radium

Table 5.1 Manchester system dosage table for planar implants and moulds (mg h per 10 Gy for radium or Ra eq sources with 0.5 mm Pt filtration).† (Data from Gibbs and Massey (1980).)

Area	Treating distance, *h* (mm)					
(cm^2)	5	10	15	20	25	30
0	32	129	289	514	804	1157
1	73	185	353	585	874	1230
2	105	230	405	646	934	1293
3	130	267	454	702	993	1353
4	152	300	499	754	1048	1409
5	174	331	540	801	1101	1465
6	191	360	579	845	1151	1517
7	207	388	614	885	1200	1570
8	222	415	647	923	1247	1620
9	239	441	677	960	1290	1668
10	254	468	707	997	1334	1717
11	268	493	737	1033	1375	1762
12	282	518	767	1069	1417	1807
13	296	542	795	1103	1456	1850
14	311	566	825	1137	1497	1893
15	326	589	852	1170	1537	1935
16	340	611	879	1202	1577	1976
17	354	632	905	1232	1611	2016
18	369	653	932	1264	1647	2057
19	383	673	957	1293	1681	2097
20	397	692	983	1323	1715	2137
22	424	728	1037	1382	1782	2213
24	450	764	1089	1442	1849	2286
26	477	796	1140	1499	1909	2363
28	503	828	1188	1553	1972	2434
30	529	859	1233	1606	2030	2506
32	554	889	1280	1660	2091	2570
34	580	922	1324	1714	2151	2637
36	603	949	1369	1769	2212	2702
38	627	982	1413	1820	2268	2767
40	651	1009	1454	1871	2324	2830
42	674	1039	1495	1922	2379	2891
44	696	1069	1534	1971	2435	2952
46	718	1096	1574	2020	2489	3011
48	740	1126	1609	2068	2542	3070
50	761	1158	1644	2115	2594	3129

† To determine the SI unit equivalent of air kerma (at 1 m) per unit system absorbed dose in $\mu Gy\ Gy^{-1}$ a conversion factor of 0.72 should be used; for example, for an area of 10 cm^2 and a treating distance of 15 mm, the figure is $707 \times 0.72 = 509$.

is related to mg Ra eq by the relationship 1 mg Ra eq is equivalent to $7.2\,\mu\text{Gy h}^{-1}$ at 1 m.

Whilst these data were primarily used for radium, the data were also used for radon by replacing mg h with mCi h. This is valid since 1 mCi of radium has approximately the same activity as 1 mg of radium (1 mg of radium = 0.989 mCi of radium or radon). The term mg h in the tables can be replaced by mCi h (MBq h) for other radionuclides. However, it is necessary to account for the fact that even though the initial activity of radon may be the same as that for radium, the number of emitted photons, and hence the dose, will not be the same with time. In order to use the tables with radon, the concept of millicurie destroyed (§1.2.2) can be employed; this enables an effective mCi h (MBq h) for each initial mCi (MBq) of radon to be determined. Suppose we have a mould of initial activity A_0, then at a time t the activity is given by equation (1.1), that is

$$A = A_0 \exp(-\lambda t).$$

Since absorbed dose is proportional to the activity and the time of application, the cumulative dose in a time T is the product of activity and time, that is

$$\begin{aligned} \int_0^T A\,\mathrm{d}t &= A_0 \int_0^T \exp(-\lambda t)\mathrm{d}t \\ &= A_0[1-\exp(-\lambda T)]\lambda^{-1} \qquad (5.1) \\ &= A_0 T_{1/2}[1-\exp(-\lambda T)]/0.693 \\ &= A_0 T_{\mathrm{ave}}[1-\exp(-\lambda T)]. \end{aligned}$$

The term $A_0[1-\exp(-\lambda T)]$ represents the activity in time T.

Equation (5.1) can also be written in terms of an effective treatment time, T_{eff}:

$$\int_0^T A\,\mathrm{d}t = A_0 T_{\mathrm{eff}} \qquad (5.2)$$

where $T_{\mathrm{eff}} = [1-\exp(-\lambda T)]/\lambda = T_{\mathrm{ave}}[1-\exp(-\lambda T)]$. The effective treatment time is defined as the time required to give the same treatment with a source of constant activity, equivalent to the initial activity A_0, as would be achieved by the decaying isotope in the actual treatment time T (Massey *et al* 1985). If the half-life of the isotope is greater than a year, then T_{eff} is approximately equivalent to the actual treatment time. In the limiting case of $T \rightarrow \infty$, then $T_{\mathrm{eff}} \rightarrow T_{\mathrm{ave}}$.

5.2.2 Principles of source distribution

Whilst the tables of Paterson and Parker give the mg h of radium required to yield an absorbed dose of 10 Gy, the dose is only achieved if source distribution rules are observed which ensure a uniformity of dose to within ±10% over the treatment surface. In the derivation of these rules, consideration was given to the equations which relate the dose at any point P on a treated area, at a distance *h* below the surface of various shaped applicators. Again, equations will be given for exposure rate but, as shown in §3.1.2, the equations for air kerma rate are almost identical.

5.2.2.1 Circular applicators. The simplest form of circular applicator is a series of point sources placed on the circumference of an applicator (figure 5.1). If the diameter of the ring is *D* cm and the treatment distance to the centre of the plane below the centre of the applicator is *h* cm, then the exposure rate due to *N* sources, each of activity *A*, is given by

$$\dot{X}_P = NA\Gamma_\delta/[(D/2)^2 + h^2]. \tag{5.3}$$

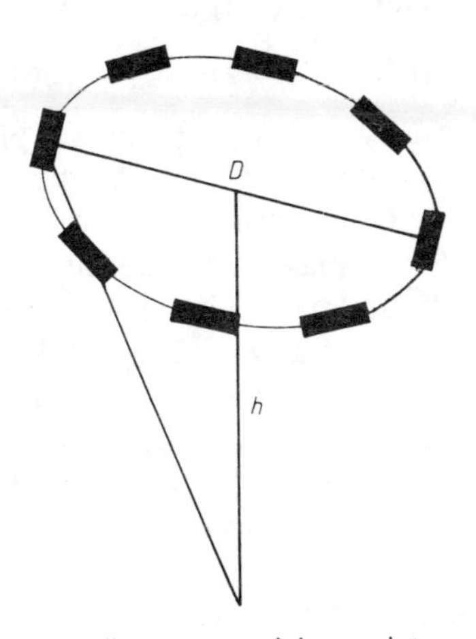

Figure 5.1 Circular applicator containing point sources on the circumference of a circle of diameter *D*.

This equation can be used to determine the exposure rate on the central axis of the mould but it will not give details of the exposure rate to off-axis points, which is essential when trying to determine the conditions for dose uniformity required in the Manchester dosage system. It is therefore necessary to consider the general equation, equation (3.9), for the exposure rate at any point on a plane at a distance *h* from a circular mould of radius r, namely

$$\dot{X}_P = A\Gamma_\delta[(r^2 + y^2 + h^2)^2 - 4r^2y^2]^{-1/2}. \qquad (5.4)$$

The variation in exposure across the diameter of the surface is a function of y; thus by differentiation we have

$$d\dot{X}_P/dy = 2y(r^2 - h^2 - y^2)A\Gamma_\delta[(r^2 + y^2 + h^2)^2 - 4r^2y^2]^{-3/2}. \qquad (5.5)$$

Using this approach Parker (1947) was able to find the maximum and minimum values of $\dot{X}_P$ for $r < h$ and $r > h$ and thus derive rules for the achievement of 'uniform' distribution in which the variation in exposure across the plane is within $\pm 10\%$.

When $r < h$, then $d\dot{X}_P/dy = 0$ if $y = 0$, hence $\dot{X}_P$ has a maximum value over the treated area for $y = 0$ and a minimum at $y = r$. For these two conditions equation (5.4) becomes

$$\dot{X}_P = A\Gamma_\delta(r^2 + h^2)^{-1} \qquad \text{for } y = 0$$

which is equivalent to equation (5.3) and

$$\dot{X}_P = A\Gamma_\delta h^{-1}(4r^2 + h^2)^{-1/2} \qquad \text{for } y = r.$$

The variation in intensity across the treatment plane can be shown to be $\pm 7.2\%$ for all rings with $(d/h) \leqslant 2$. Under the criteria used this means that the intensity is 'uniform' over the treatment plane.

However, for $r > h$ the maximum intensity over the treated area occurs at $y = (r^2 - h^2)^{1/2}$ and the minimum at either $y = 0$ for $r > (2)^{1/2}h$ or at $y = r$ for $r < (2)^{1/2}h$. For $r < (2)^{1/2}h$ the variation in intensity decreases as r/h increases, the greatest variation being $\pm 5\%$ for $r = h$ and is at least $\pm 3\%$ at $r = (2)^{1/2}h$: thus all such circles in the range $2 < d/h < 3$ satisfy the 'uniformity' criteria. In fact Mayneord showed the most uniform distribution occurs for the 'ideal' circle where $d/h = 2(2)^{1/2} = 2.83$.

If $r > (2)^{1/2}h$, the minimum intensity is at the centre of the treated area and this situation can lead to the $\pm 10\%$ uniformity criteria not being kept by the use of a single ring. To overcome this decrease in the central dose it was proposed that additional activity be added to the centre of the mould. Suppose the amount of activity added at the centre is dA, then the dose rate at any point on the surface will be

$$dA\Gamma_\delta(h^2 + y^2)^{-1} + A\Gamma_\delta[(r^2 + y^2 + h^2) - 4r^2y^2]^{-1/2}. \qquad (5.6)$$

Equation (5.6) can now be used to derive the quantity dA which will make the dose rate almost constant over the range $y = 0$ to $y = r$, the point of minimum dose rate. Thus for equality at $y=0$ and $y=r$,

$$\mathrm{d}A\Gamma_\delta h^{-2} + A\Gamma_\delta(r^2+h^2)^{-1} = \mathrm{d}A\Gamma_\delta(h^2+r^2)^{-1} + A\Gamma_\delta[h(4r^2+h^2)]^{-1}.$$

Similarly, the activity required to achieve equality at $y=0$ and at the point of maximum intensity when $y=(r^2-h^2)^{1/2}$ can be determined. In these situations the value to be added to the centre is 5.3% and 6.7%, respectively, of the activity in the ring. Using this approach the percentage of activity to be placed at the centre was set at 5% for $d/h=3$ to less than 6. For a value of $d/h \geqslant 6$, however, the variation in dose across the plane increases and it is therefore necessary to add a further ring of sources at half the previous diameter. These considerations led to the distribution rules for circular applicators as follows (modified from data of Parker (1947)).

(i) The distance between the *active* ends should be less than the treating distance h.
(ii) Circular applicators of diameter d and at treating distance of h should have the activity distributed in an outer circle, an inner circle of diameter $d/2$ and a central spot with the percentages as indicated in table 5.2.
(iii) There must be a minimum of six sources to achieve a circular arrangement.
(iv) Ellipses of small eccentricity may be used and considered as circles.

Table 5.2 Distribution of source activity in circular applicators.

	$d/h<3$	$3<d/h<6$	$6<d/h<7.5$	$7.5<d/h<10$	$d/h>10$
Outer circle	100	95	80	75	70
Inner circle	0	0	17	22	27
Central spot	0	5	3	3	3

5.2.2.2 Plane rectangular applicators. To study dose uniformity on a plane due to a rectangle of sources, Parker constructed isodose curves using Sievert tables. The rectangle was considered to be made up of four line sources and the dose to any point over the treated area was obtained by summing the contribution due to each line source to points on the treatment surface. It was found that when the side of the square had a length equal to twice the treating distance, then the uniformity of distribution was similar to that obtained with an 'ideal' circle. 'Uniformity' of dose to satisfy the ±10% criteria can also be achieved for a rectangle ab, with $b > a$, when $a = 2h$ and $b < 4a$, if the dose in the corners is excluded. For larger rectangles, it was found necessary to add extra lines parallel to the longer side at a separation of $2h$ in order to achieve uniformity of dose, but the source activity per cm required was only 1/2 of the periphery activity per cm for one added line and 2/3 for two or more lines. This rule was found to be applicable for rectangles whose ratio of sides is 4 : 1—although as elongation increases it is found that the number of mg h for 10 Gy needs to be increased.

In a comparison of the dose received at a point per mg h from circular and rectangular applicators of the same area, Parker found that circles and squares have almost the same dose and that rectangles with an elongation of 2 : 1 only differ by up to 6%. From this follows the important fact that, provided the sources are arranged in accordance with the distribution rules, then the dose rate per mg is dependent on the area of the applicator rather than the shape. It was this conclusion which led to the Paterson–Parker dosage tables which relate area to the number of mg h for 10 Gy for various treating distances. For rectangular applicators the distribution rules are as follows (modified from data of Parker (1947)).

(i) The distance between *active* ends should not be greater than $2h$.
(ii) For rectangles and/or squares of area ab, where $b > a$, sources should be placed around the periphery and possibly some in lines parallel to b, the number of lines being such that the area is divided into strips of width $2h$.
(iii) If the activity per unit length on the periphery is ρ, then (a) for $a < 2h$ all the actvity should be placed at the periphery, (b) for $2h < a < 4h$ one added line is required whose linear density is $1/2\rho$, (c) for $a > 4h$ two or more lines are added whose linear density is $2/3\rho$.
(iv) The number of mg h for 10 Gy must be increased by a correction factor to account for any elongation as given in the table below

Elongation factor $b:a$	1.5 : 1	2 : 1	3 : 1	4 : 1
Multiplying factor on mg h	1.025	1.05	1.09	1.12

Whilst these rules apply strictly to flat, circular or square applicators (where the treating area is equal to the applicator area), they also apply to curved surfaces. To calculate the amount of nuclide required, the area to be used in the tables must be (a) the area of the applicator, regardless of the area treated, for *concave* surfaces and (b) the area to be treated, even though the sources are to be spread over the larger applicator area, for *convex* surfaces. It must be emphasised that the dose table (table 5.1) only achieves the $\pm 10\%$ dose uniformity when used in conjunction with the distribution rules given in §§5.2.2.1 and 5.2.2.2.

5.2.2.3 Cylindrical moulds. The clinical requirement of these moulds is to treat the lesion by surrounding it with an arrangement of sources. Parker (1947) and Spiers (1947) both considered the variations in dose within and around such a cylinder. Using a mathematical approach, Spiers firstly considered the 'intensity' at a point P on the end plane of the cylinder. Since the whole distribution is symmetrical about the axis of the cylinder, the intensity at any point on a coaxial circle through P will be the same as that at P. Now the surface of a cylinder can be regarded as being constructed of a large number of rings of thickness dh. Therefore, the intensity at P due to a cylinder of radius r can be obtained by integrating equation (3.9), the exposure rate expression for a ring source, over the whole length of the cylinder, H; that is

$$\dot{X}_{\mathrm{P}} = A\Gamma_{\delta} \int_0^h \mathrm{d}h\{[(r+y)^2 + h^2][(r-y)^2 + h^2]\}^{-1/2} \qquad (5.7)$$

and substituting $z = h/(r-y)$, equation (5.7) becomes

$$\dot{X}_{\mathrm{P}} = [(A\Gamma_{\delta})/(r+y)] \int_0^z \mathrm{d}z[(1+z^2)(1+c^2z^2)]^{-1/2} \qquad (5.8)$$

where $c = (r-y)/(r+y)$.

To evaluate the elliptical integral in equation (5.8) standard tables can be used. The dose at any point within the cylindrical mould may be determined by considering the point to be on the end of two small cylinders of length h_1 and h_2, where $H = h_1 + h_2$. The intensity at the point can thus be represented as the sum of two integrals. This method enabled Spiers to confirm the rules devised by Parker, who considered the same problem by postulating that a cylindrical applicator could be considered as a special case of a rectangular applicator in which opposite sides of the rectangle coincided. From these considerations, rules were derived in terms of d/D and L/D, where d is the diameter of the surface, D is the diameter of the source cylinder and L is its length.

Cylindrical moulds can be constructed in a number of ways. Parker (1947) defined Type 1 moulds as those in which the radium is mounted on the surface of the cylinder as a series of coaxial rings, whilst Type 2 moulds consist of a series of equidistant lines of radium parallel to the axis of the cylinder. To determine the amount of radium or mg Ra eq to by used, the number of mg h required to deliver 10 Gy at the centre of a ring of diameter D equal to that of the proposed treating cylinder must be found using the formula

$$\text{mg h per 10 Gy for 0.5 m Pt} = 30 \times 1.08D^2. \tag{5.9}$$

In order to find the number of mg h required to give a dose of 10 Gy on the skin surface within the cylinder, then an appropriate multiplication factor must be used (table 5.3), this factor being obtained by taking into account the ratio of the length L to the cylinder diameter, D, and the ratio of the skin surface diameter, d, and the cylinder diameter, D.

Table 5.3 Factors for Type 1 and Type 2 cylinders. (Data modified from Meredith (1967).)

	d/D							
L/D	0.0	0.1	0.2	0.3	0.4	0.5	0.6	0.7
0.0	1.00	0.98	0.96	0.91	0.84	0.75	0.64	0.51
0.1	1.01	1.00	0.98	0.93	0.86	0.78	0.68	0.56
0.2	1.06	1.04	1.02	0.98	0.92	0.84	0.75	0.64
0.3	1.12	1.11	1.09	1.04	1.00	0.93	0.83	0.73
0.4	1.18	1.17	1.15	1.11	1.07	1.00	0.90	0.80
0.5	1.24	1.23	1.21	1.17	1.13	1.07	0.97	0.87
0.6	1.31	1.30	1.28	1.24	1.20	1.12	1.04	0.94
0.7	1.38	1.37	1.35	1.31	1.27	1.20	1.10	1.01
0.8	1.45	1.43	1.41	1.38	1.34	1.28	1.18	1.08
0.9	1.52	1.50	1.48	1.45	1.41	1.35	1.25	1.15
1.0	1.60	1.58	1.56	1.53	1.49	1.43	1.33	1.21
1.1	1.66	1.65	1.63	1.60	1.56	1.49	1.39	1.27
1.2	1.73	1.71	1.69	1.66	1.62	1.55	1.45	1.33
1.3	1.80	1.78	1.76	1.73	1.68	1.62	1.52	1.40
1.4	1.86	1.85	1.83	1.79	1.74	1.68	1.58	1.46
1.5	1.94	1.92	1.90	1.86	1.81	1.74	1.64	1.53
1.6	2.00	1.98	1.96	1.92	1.87	1.80	1.69	1.57
1.7	2.08	2.06	2.04	2.00	1.95	1.87	1.76	1.63
1.8	2.16	2.14	2.12	2.07	2.02	1.94	1.83	1.69
1.9	2.24	2.22	2.19	2.15	2.10	2.02	1.90	1.76
2.0	2.31	2.29	2.27	2.23	2.18	2.09	1.97	1.82

Having determined the amount of activity required, then the distribution rules are as follows (modified from data of Parker (1947)).

(i) Distance between coaxial rings must be twice the 'distance' of the applicator, i.e. $(D - d)$.
(ii) The distribution of sources depends on the number of rings required (table 5.4).
(iii) The ring is preferably composed of a continuous chain of needles or tubes.
(iv) If the sources are mounted in a series of straight lines, the intervals around the circumference must be equal to the thickness of the applicator $(D - d)$. A ring must be added at each end and the activity per cm of the straight lines is 2/3 the activity per cm in the ends.

Table 5.4 Distribution of sources.

Number of rings	Per cent of total in outer rings	Per cent of total in inner rings
2	50 + 50	—
3	40 + 40	20
4	30 + 30	20 + 20
5	25 + 25	16.7 + 16.7 + 16.7

5.2.3 Surface mould depth doses

Whilst the dosage tables indicate the number of mg h required to give 10 Gy to the surface of the skin, it is often necessary to know the absorbed dose at some depth below the skin. This can readily be determined from the dose tables, as shown below.

Example 5.1

Consider a mould, 10 cm^2 in area. For a treating distance of $h = 0.5$ cm, 254 mg h are required to give a surface dose of 10 Gy. Making the assumption that the sources are separated from the skin surface by tissue equivalent material, then from table 5.1 we find that 468 mg h are required to give a dose of 10 Gy at a distance of 1.0 cm from the sources, i.e. 0.5 cm below the skin. Since the dose of 10 Gy is prescribed for the skin, the mould is made with 254 mg h of radium, thus the plane 0.5 cm below the skin will only receive a dose of $(254/468) \times 10$ Gy $= 5.43$ Gy; i.e. 54.3% of the skin dose.

Table 5.5 Percentage depth dose (DD) for various treating distances h.

Treating distance, h(mm)	mg h per 10 Gy for 10 cm^2 circle at specified distance										
	$h+0$ (mm)	$h+5$ (mm)	%DD at 5 mm	$h+10$ (mm)	%DD at 10 mm	$h+15$ (mm)	%DD at 15 mm	$h+20$ (mm)	%DD at 20 mm	$h+30$ (mm)	%DD at 30 mm
5	254	468	54.3	707	35.9	997	25.5	1334	19.0	2182	11.6
10	467	707	66.2	997	46.9	1334	35.1	1717	27.3	2700	17.3
15	707	997	70.9	1334	53.0	1717	41.2	2182	32.4	3226	21.9
20	997	1334	74.7	1717	58.1	2182	45.7	2700	36.9	3866	25.8
25	1334	1717	77.7	2182	61.1	2700	49.4	3226	41.4	—	—
30	1717	2182	78.7	2700	63.6	3226	53.2	3866	44.4	—	—

In general we can say that the percentage depth dose can be written

$$\frac{\text{mg h per 10 Gy at treating distance}}{\text{mg h per 10 Gy at (treating distance + depth)}} \times 100. \qquad (5.10)$$

Using this relationship, the variation in percentage depth dose for different treating distances is shown in table 5.5. It can readily be seen that the percentage depth dose at 0.5 cm increases as the treating distance increases. This effect is *mainly* due to the inverse square law effect. If, therefore, it is necessary to treat a thin superficial lesion whilst at the same time avoiding irradiating underlying structures, then we require a mould with a small treating distance since this gives a lower percentage depth dose. Conversely, for thicker lesions, in order that the dose gradient across the volume is small, we need to have a greater percentage depth dose, thus h must be increased, although this does mean using either a longer treatment time or more source activity to achieve the same skin dose. These features make moulds particularly attractive in the treatment of lesions, for example on the dorsum of the hand, where underlying structures need to be minimally irradiated.

Account also has to be taken of the percentage depth dose when constructing a double or sandwich mould. It is often advantageous to sandwich a growth between two moulds, thereby ensuring a considerable degree of homogeneity of dose throughout the block of tissue, since the fall in intensity due to one mould is balanced by the other. Care in construction is necessary to ensure that the two moulds are parallel to one another.

5.2.4 Radium substitutes

The Paterson–Parker dosage system was designed for use with radium and radon. With the decline in the use of these isotopes the tables are being increasingly used with other radionuclides, such as caesium-137, gold-198 and iridium-192. This is equally valid provided due allowance is made for the different exposure rate constants, filtration and, where necessary, the half-life of the radio-isotope. It is not, however, justified to use the Paterson–Parker tables for gamma-ray emitters with photon energy of less than 0.2 MeV or where the alpha- and beta-particles are not adequately filtered.

As discussed in Chapter 4, the source activity can be defined in terms of an effective equivalent mass of radium, mg Ra eq. Using this method of source activity specification, the radium dosage tables can be used with other radionuclides—radium substitutes. It must be emphasised, however, that the mg Ra eq of a source is not necessarily the ratio of the exposure rate constant of the radium substitute to radium; account must also be taken of the differences in filtration and source geometry. For example: for a caesium source, the ratio of exposure/air kerma rate constants suggests that the activity of caesium-137 required for 1 mg Ra eq would be $0.825/0.3275 = 2.52$ mCi of caesium-137 per mg of radium-226, or $7.2/0.0773 = 93.1$ MBq of caesium-137 per mg of radium-226. In reality, however, the effective mg Ra eq for a caesium-137 tube encased in 0.5 mm platinum is $(7.2/0.0773)\exp(-\mu \times 0.05) =$ 99.2 MBq (2.68 mCi) of caesium-137, since $\mu = 1.25\ \text{cm}^{-1}$ (Breitman 1974). If the tube is constructed with stainless steel walls, where $\mu = 0.221\ \text{cm}^{-1}$, the equivalent activity for 1 mg Ra eq would be $(7.2/0.0773)\exp(-0.221 \times 0.05) = 94.2$ MBq (2.55 mCi) of caesium-137.

These differences are overcome by defining source strength in terms of air kerma/exposure rate at 1 m, as discussed in Chapter 4. The mg Ra eq activity of the nuclide can then be obtained by dividing the measured exposure rate by the air kerma/exposure rate of a 1 mg point source of radium at the source distance filtered by 0.5 mm of platinum, i.e. $7.2\ \mu\text{Gy h}^{-1}$ or $0.825\ \text{mR h}^{-1}$. If, however, the sources are specified in terms of equivalent activity derived from an air kerma/exposure measurement, then to convert to mg Ra eq it is necessary to use the value of the air kerma/exposure rate factor used by the suppliers in determining the equivalent activity. For example, gold grains having an equivalent activity A_e will have a mg Ra eq of $A(\text{mg Ra eq}) = (0.0555 A_e)/7.2$ for A_e in MBq, or $A(\text{mg Ra eq}) = (0.235 A_e)/0.825$ for A_e in mCi.

When short-lived radionuclides are used as a substitute for radon, other factors have to be considered when using the radium tables. We know that in order to use the Paterson–Parker tables it is necessary to determine the cumulative dose in mg h Ra eq per initial activity of radionuclide. This is given by equation (5.2); thus it is necessary to calculate T_{eff} for the isotopes used. Values of T_{eff} for isotopes commonly used in brachytherapy are given in table 5.6. They are found by evaluating equation (5.1).

By using the source activity specified in terms of mg Ra eq and T_{eff}, the cumulative dose in mg h Ra eq per initial activity can be determined as shown in table 5.7 for initial activities of gold-198, quoted in both mCi and MBq.

Due to the difference in half-life between radon and gold, consideration has also to be given to the difference in initial activity required to achieve the same absorbed dose. In total decay the absorbed dose due

Table 5.6 Effective treatment time for various radionuclides.

Actual treatment time (days)	T_{eff} for various nuclides (hours)			
	I-125 $T_{1/2} = 60.1$ days	Ir-192 $T_{1/2} = 74.0$ days	Au-198 $T_{1/2} = 64.7$ hours	Rn-222 $T_{1/2} = 91.8$ hours
0.5	12.0	12.0	11.3	11.5
1.0	23.8	23.9	21.2	22.0
1.5	35.7	35.8	29.9	31.5
2.0	47.5	47.6	37.5	40.3
2.5	59.1	59.3	44.3	48.3
3.0	70.8	71.0	50.2	55.5
3.5	82.3	82.6	55.4	62.2
4.0	93.8	94.2	60.0	68.3
4.5	105.3	105.8	64.0	73.8
5.0	116.6	117.2	67.5	78.9
5.5	127.9	128.7	70.7	83.6
6.0	139.1	140.0	73.4	87.8
6.5	150.3	151.4	75.8	91.7
7.0	161.4	162.6	77.9	95.2
7.5	172.4	173.8	79.8	98.4
8.0	183.4	185.0	81.4	101.4
8.5	194.3	196.1	82.9	104.1
9.0	205.2	207.2	84.1	106.5
9.5	216.0	218.2	85.2	108.8
10.0	226.7	229.1	86.2	110.8
11.0	247.9	250.9	87.8	114.4
12.0	269.0	272.4	89.1	117.4
20.0	428.7	437.7	92.8	128.9
∞	2080.9	2562.2	93.4	132.4

to 1 mCi of radon would be equivalent to that received from $(7.2/0.0555)(132.4/93.4) = 184$ MBq of gold, or $(0.825/0.235)(132.4/93.4) = 4.98$ mCi of gold. This value is in approximate agreement with the proposed value of 4.9 mCi given in the Radiochemical Centre Technical Publications 66/8, 68/8, 69/6 and 71/7 and cited by Binks (1981), the differences being due to the use of later figures for half-life and exposure rate constant. However, this relationship cannot be assumed to be constant for all time. The variation in initial activities required to give the same dose as a function of time is given by

$$(A_0 \Gamma_\delta T_{eff})_{Rn} = (A_e \Gamma_\delta T_{eff})_{Au}.$$

Using this expression, it is possible to derive the initial equivalent activity of gold required to achieve the same dose as radon with time. This is

Table 5.7 Cumulative dose in mg h Ra eq per initial equivalent activity of gold and equivalent activity of gold per unit activity of radon.

Days	Cumulative dose in mg h Ra eq per initial mCi of gold	Cumulative dose in mg h Ra eq per initial MBq of gold	Equivalent activity of gold per unit activity of radon
0.5	3.21	0.087	3.58
1.0	6.03	0.163	3.64
1.5	8.51	0.230	3.70
2.0	10.69	0.289	3.77
2.5	12.61	0.341	3.83
3.0	14.29	0.386	3.89
3.5	15.78	0.426	3.94
4.0	17.08	0.462	4.00
4.5	18.22	0.493	4.05
5.0	19.24	0.520	4.10
5.5	20.12	0.544	4.15
6.0	20.90	0.565	4.20
6.5	21.59	0.584	4.24
7.0	22.19	0.600	4.29
7.5	22.72	0.614	4.33
8.0	23.19	0.627	4.37
8.5	23.60	0.638	4.41
9.0	23.96	0.648	4.44
9.5	24.28	0.656	4.48
10.0	24.56	0.664	4.51
11.0	25.02	0.676	4.57
12.0	25.37	0.686	4.63
20.0	26.43	0.714	4.88
∞	26.59	0.719	4.98

Table 5.8 Radium-226 or caesium-137 F-type sources.

Type code	Source strength mg Ra eq	Source strength Air kerma rate (μGyh^{-1} at 1 m)	External length (mm)	Active length (mm)
F-1	1.5	10.8	7.5	5
F-2	2.5	18.0	12.5	10
F-3	3.0	21.6	7.5	5
F-4	5.0	36.0	12.5	10

shown in table 5.8 and is based on the exposure/air kerma rate factors used by Amersham International plc.

In addition to these factors a biological factor needs to be considered when using gold, since according to Paterson, the use of gold-198 should be 10% more effective due to its shorter half-life. Thus, in order to achieve the same biological effect, the Radiochemical Centre suggest that a 1 mCi radon seed should be regarded as biologically equivalent to a 4.5 mCi gold seed. This increase in biological effectiveness can be deduced from the data of Paterson or from the TDF equation of Orton (1974).

5.2.5 Clinical examples

Initially, a considerable range of different radium sources was available for moulds. Many of these sources were replaced by equivalent caesium-137 tubes and needles, but with the decline in the use of this modality of treatment the range of sources available for use on moulds is severely limited. To illustrate the calculation of applicator treatment, sources have been chosen which were formerly commercially available as radium or caesium F-type sources. These are given in table 5.8.

It must be remembered that the Paterson–Parker tables can be used with other isotopes expressed in terms of mg Ra eq, although in practice it may not be possible to comply with the distribution rules. This is the case when iridium-192 wire of constant linear activity is used.

5.2.5.1 Simple ring mould. Eight 3.0 mg F-3 type tubes are placed on the circumference of a 4.0 cm diameter circular applicator (figure 5.2(*a*)). What is the exposure rate and air kerma rate in air at a distance of 1.5 cm from the plane of the applicator, the exposure rate constant for radium-226 being 0.825 mR h^{-1} at 1 m? The air kerma rate at 1 m from a 1 mg Ra eq source filtered by 0.5 mm Pt is 7.2 μGy h^{-1}. Use table 1.1 to determine the absorbed dose in tissue for both answers.

Since each source may be considered to be a point source, then, assuming constant filtration, the exposure rate is given by equation (5.3):

$$\dot{X}_P = 8 \times 3.0 \times 0.825(100/1)^2/1000[(4/2)^2 + 1.5^2]$$
$$= 31.68 \text{ R h}^{-1}.$$

The air kerma rate, $\dot{K}_a$, is obtained using a similar equation:

$$\dot{K}_a = 8 \times 3.0 \times 7.2(100/1)^2/[(4/2)^2 + 1.5^2]$$
$$= 27.65 \text{ cGy h}^{-1}.$$

Using the cGy factor and the value of the ratio of the mass energy absorption coefficients of tissue to air in table 1.1, the dose rate in tissue, $\dot{D}_m$, is given by

$$\dot{D}_m = 31.68 \times 0.971 = 30.76 \text{ cGy h}^{-1}$$

or

$$\dot{D}_m = 27.65 \times 1.112 = 30.74 \text{ cGy h}^{-1}.$$

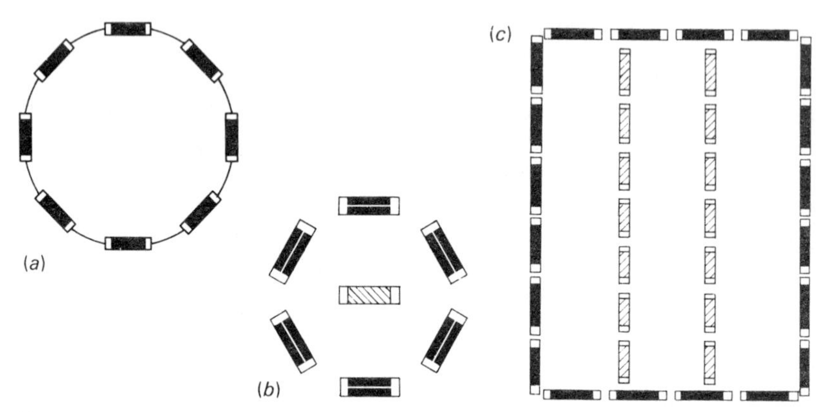

Figure 5.2 (*a*) Circular mould with a treating distance of 1.5 cm containing eight 3.0 mg Ra eq caesium-137 F-3 type sources. (*b*) Arrangement of sources on a circular mould for treating a lesion of 4.0 cm diameter on the dorsum of the hand, the treating distance being 1.0 cm. (*c*) Arrangement of sources in a rectangular mould.

5.2.5.2 Circular mould. A lesion on the dorsum of the hand is to be treated with a 4.0 cm diameter applicator at a treating distance of 1 cm. The applicator is to be worn for approximately 8 hours per day to give a prescribed dose of 60 Gy in 6 days. How many mg of radium or radium-equivalent are required, what is the ideal distribution of the sources and what is the dose 0.5 cm below the surface of the skin?

Area to be treated = 12.6 cm^2, total treatment time = 48 hours: from table 5.1, mg Ra eq h for 10 Gy for $h = 1.0$ is 532; mg Ra eq h of radium for 10 Gy for $h = 1.5$ is 784. For treatment of 60 Gy total mg h required = 532×6 mg h. Total number of mg required = $532 \times 6/48 = 66.5$ mg.

For this mould $d/h = 4$. Thus, from table 5.2, 95% of the activity (63.2 mg) must be placed on the periphery and 5% (3.3 mg) at the centre. The practical limitation of the sources means that on the circumference twelve F-4 tubes are used (total activity = 60 mg) and at the central spot one F-3 tube is used (activity = 3 mg), as shown in figure 5.2(*b*).

In order to achieve the dose prescribed with these sources the applicator must be worn for $532 \times 6/63 = 50.7$ hours; i.e. it must be worn for 8 hours 26 mins each day for 6 days. The dose received to the plane 0.5 cm below the skin surface is given by equation (5.10):

$$532 \times 60/784 = 40.7 \text{ Gy}.$$

5.2.5.3 Rectangular mould. A superficial lesion of the forehead covering $4 \times 6 \text{ cm}^2$ is to be treated with a mould $6 \times 8 \text{ cm}^2$ at a treating distance of 1 cm. The mould is to be worn for 8 hours a day for 6 days and the prescribed dose is 60 Gy. Calculate the number of mg h of radium or radium-equivalent required and the optimum distribution of sources.

(i) *Plan.* Area = 48 cm^2, elongation factor = 1.025; from table 5.1, for $h = 1$ cm, mg h for 10 Gy = 1126. For this elongated mould the number of mg h required for 60 Gy = $1126 \times 1.025 \times 6 = 6925$ mg h. The mould is to be worn for approximately 48 hours, therefore we require 144 mg of radium or radium-equivalent sources.

(ii) *Source distribution.* The distance between the lines = $2h$ = 2 cm, thus we need two added lines parallel to the long side, each with a linear density 2/3 of the linear activity of the periphery. If ρ is the linear activity of the outer sides in mg cm^{-1}, then

$$(8 + 8 + 6 + 6)\rho + (2/3)\rho(8 + 8) = 144$$
$$38.7\rho = 144$$

therefore

$$\rho = 3.73 \text{ mg cm}^{-1}.$$

Thus each long side should contain approximately 29.8 mg, each short side should contain approximately 22.3 mg and each central line approximately 19.8 mg. The practical distribution of sources, shown in figure 5.2(*c*), is therefore as given in table 5.9. The total number of mg = (30 + 30) + (20 + 20) + (21 + 21) = 142 mg. Since the total activity is less than required, the overall time required is $1126 \times 1.025 \times 6/142 = 48.77$ hours, thus the mould is to be worn for 8 hours 8 mins per day for 6 days.

Table 5.9 Distribution of sources for rectangular mould.

Line	Tube type	Number of sources	Total activity (mg)	Gaps between sources (mm)
Long side	F-4	2 rows of 6	2 × 30	1
Short side	F-4	2 rows of 4	2 × 20	2
Bar	F-3	2 rows of 7	2 × 21	4

5.2.5.4 Gold grain elastoplast mould. A gold grain elastoplast mould is to be used to treat a lesion on the leg to a prescribed dose of 55 Gy in 5 days at a treating distance of 0.5 cm. If the diameter of the applicators is 2.5 cm, what activity of gold is required and how are the seeds to be distributed? (Area of applicator = 4.9 cm^2, $d/h = 5$.)

From table 5.1, mg Ra eq for 10 Gy of radium = 172, therefore the number of mg h for 50 Gy = 860 mg Ra eq h. Since the effective treatment time for 5 days is 67.5 hours (table 5.6), and the cumulative dose per initial 1 MBq (1 mCi) of gold-198 for 5 days is 0.52 (19.24) mg Ra eq (table 5.7), then the total activity of gold required is 860/0.52 = 1654 MBq or 860/19.24 = 44.7 mCi.

For a mould where $d/h = 5$, we require 95% of the total activity on the periphery and the remaining 5% at the centre. In order to comply with the rule of not having more than a distance of $2h$ between the active ends, we must have at least 16 seeds on the periphery, which is 7.85 cm long, and the maximum number which can be accommodated on the mould is 30 since each seed is 2.5 mm in length. Therefore, using 85 MBq (2.3 mCi) seeds we place 19 seeds around the periphery and 1 seed as a central spot source. The total activity used, therefore, is 1700 MBq (46.0 mCi): thus the treatment time must be reduced by 1654/1700 = 2.8% to 116 hours 45 mins.

When source strength is specified in air kerma rate at 1 m, the above example can be calculated as follows.

For 50 Gy the total activity required is 860 mg Ra eq h. Since 1 mg Ra eq source has a source strength of 7.2 μGy h^{-1} at 1 m, then for a dose of 50 Gy we require 860 × 7.2 μGy h^{-1} h at 1 m = 6192 μGy at 1 m.

For a five-day treatment the effective treatment time is 67.5 hours, therefore the total source strength required in μGy h^{-1} at 1 m for 50 Gy is 6192/67.5 = 91.73 μGy h^{-1} at 1 m. To achieve the distribution rules we require 20 seeds, thus each seed should have a strength of 4.59 μGy h^{-1}

at 1 m. If the sources arrive with a strength of 4.7 μGy h^{-1} at 1 m, then the treatment time has to be reduced by 2.5% to 117 hours.

5.2.5.5 Sandwich mould. A lip lesion, 1.5×2.5 cm^2, is to be treated using a sandwich mould, figure 5.3(*a*), the lip being 1 cm thick. The mould is to be worn for approximately 6 hours daily for 8 days and the desired doses are 55 Gy on the skin surface and 75 Gy on the mucosa. How many mg of radium or radium-equivalent sources are to be placed on each mould?

To determine the area of each mould consideration has to be given to the curvature on the lip (a radius of curvature of 8 cm) and also the depth limitation on the inside of the lip. For the *inner* mould all surfaces are *concave*, thus for the purposes of calculation the area covered by the sources on the mould must be used. However, the *outer* mould is treating a series of *convex* surfaces and it is these areas which must be used for calculation, the sources being spread over the larger area of the mould. The *inner* mould has a treating distance of 0.5 cm, whilst the *outer* mould is 1 cm from the skin surface (figure 5.3(*b*)). For adequate coverage the area to be covered in the middle of the lip is 2.5×4.5 cm^2.

(i) *Inner mould.* Area of mould, therefore, is 2.5×4.0 cm^2 and $h = 0.5$. The characteristics of the mould are given in table 5.10.

(ii) *Outer mould.* The characteristics are as given in table 5.11.

If the inner mould contributes X Gy to the mucosa and the outer mould Y Gy to the skin, then two simultaneous equations can be derived to achieve the prescribed dose, that is

$$X + 0.546Y = 75$$
$$0.359X + Y = 55$$

Table 5.10 Inner mould characteristics.

Parameter	Mucosa	Position Centre of lip	Skin
Distance from sources (cm)	0.5	1.0	1.5
Number of mg h per 10 Gy	254	468	707
Elongation factor	1.02	1.02	1.02
Required mgh for 10 Gy	259	477	721
Percentage depth dose†	100	54.3	35.9

† Derived using equation (5.10) in §5.2.3.

Table 5.11 Outer mould characteristics.

Parameter	Skin	Position Centre of lip	Mucosa
Effective area (cm^2)†	3.5 × 5.0 = 17.5	3.5 × 4.7 = 16.5	3.5 × 4.4 = 15.4
Distance from sources (cm)	1.0	1.5	2.0
Number of mg h per 10 Gy	643	892	1183
Elongation factor	1.015	1.015	1.010
Required mg h per 10 Gy	653	905	1195
Percentage depth dose	100	72.2	54.6

† The effective area is found for the *convex* surface by multiplying the height of the mould, 3.5 cm, by the length in each plane, assuming a radius of curvature of 8 cm, as shown in figure 5.3(*c*).

therefore

$$0.804X = 44.97$$
$$X = 55.93 \text{ Gy}$$

and thus $Y = 34.93$ Gy.

In order to give a combined dose of 75 Gy to the mucosa and 55 Gy to the skin, the *inner* mould alone must contribute 55.93 Gy to the mucosa and the *outer* mould alone must contribute 34.93 Gy to the skin. Using these values the two moulds are treated separately for calculation.

For the inner mould, the number of mg h required for a dose of 55.93 Gy to the mucosa $= 254 \times 1.02 \times 5.593 = 1.449.0$ mg h. To give this dose in 48 hours requires 30.2 mg radium or radium-equivalent sources. Since the applicator has an area of $2.5 \times 4.0\ \text{cm}^2$ and $h = 0.5$, we must use two 'bars' parallel to the longer side, each with 2/3 the linear activity, ρ, of the periphery. Therefore to find ρ we calculate

$$(2.5 + 2.5 + 4 + 4)\rho + 2 \times 4(2/3)\rho = 30.2$$
$$18.33\rho = 30.2$$
$$\rho = 1.65\ \text{mg cm}^{-1}.$$

Thus, each long side should contain approximately 6.6 mg, each short side approximately 4.1 mg and each 'bar' approximately 4.4 mg.

The inner mould is therefore constructed with the sources as shown in table 5.12.

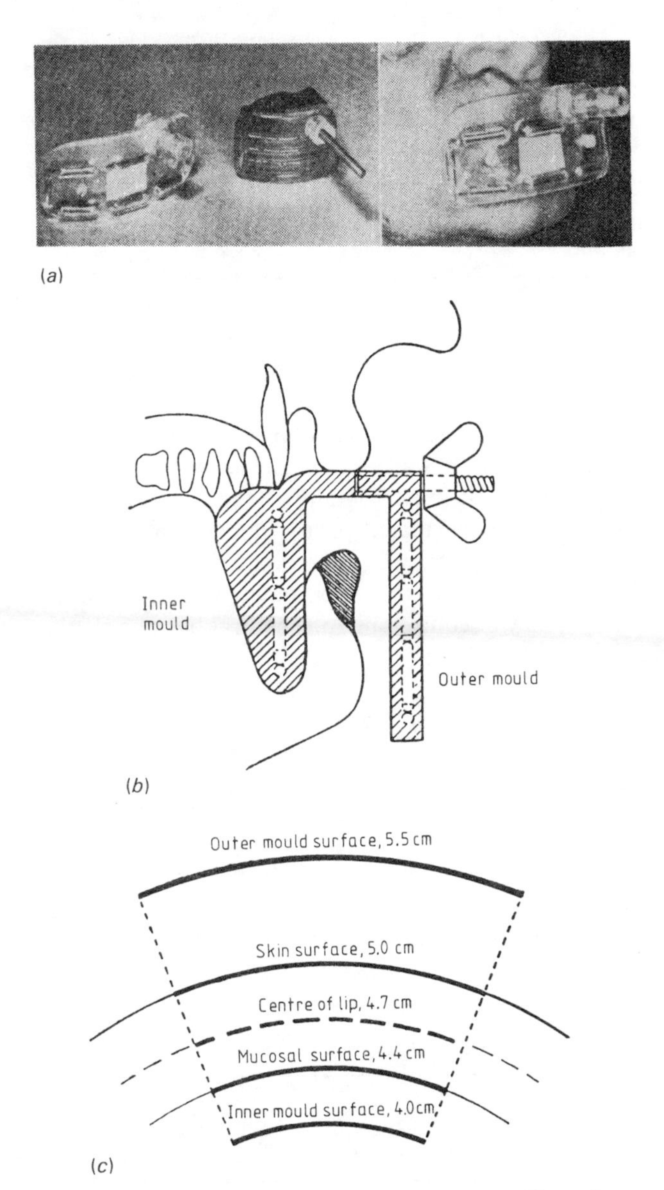

Figure 5.3 (*a*) Construction of sandwich lip mould. (From Paterson (1963).) (*b*) Diagram of sandwich mould for treating lip lesion. (Reproduced from Meredith (1967) by permission of Churchill Livingstone Ltd, Edinburgh.) (*c*) Diagram of mould radii of curvature.

Table 5.12 Inner mould source distribution.

Line	Tube type	Number of sources	Total activity (mg)	Gaps between sources (mm)
Long side	F-1	2 rows of 4	2 × 6	5
Short side	F-1	2 rows of 3	2 × 4.5	1
Bar	F-1	2 rows of 3	2 × 4.5	9

The total activity in the inner mould is therefore 30 mg Ra eq, thus the time taken to deliver a dose of 55.93 Gy is $1449.0/30 = 48.3$ hours.

For the outer mould the number of mg h required to deliver a dose of $34.93 \text{ Gy} = 643 \times 1.015 \times 3.493 = 2281$ mg h. Since the mould is to be worn for 48.3 hours, it must contain 47.2 mg.

The distribution is calculated on the dimensions of the treated skin area, i.e. $3.5 \times 5.0 \text{ cm}^2$, but the sources are distributed over an area of $3.5 \times 5.5 \text{ cm}^2$. Since $h = 1$ cm, the distance between each line must not exceed 2 cm, thus we require only one central bar at half the peripheral linear activity, ρ. We have

$$(5 + 5 + 3.5 + 3.5)\rho + 5 \times (1/2)\rho = 47.2$$
$$19.5\rho = 47.2$$
$$\rho = 2.42 \text{ mg cm}^{-1}.$$

Thus each long line should could 12.1 mg, each short line should contain 8.5 mg and the central bar 6.0 mg.

For the outer mould the sources available are arranged as shown in table 5.13.

Table 5.13 Outer mould source distribution

Line	Tube type	Number of sources	Total activity (mg)	Gaps between sources (mm)
Long side	F-3	2 rows of 4	2 × 12	8
Short side	F-3	2 rows of 3	2 × 9	4
Bar	F-1	1 row of 4	1 × 6	7

Since the total activity differs from that calculated, the treatment times change—thus a check calculation must be made (see table 5.14).

Table 5.14 Check calculations for treatment times.

	Position					
	Inner mould, 30 mg			Outer mould, 48 mg		
Parameter	Mucosa	Centre	Skin	Skin	Centre	Mucosa
Number of mg h per 10 Gy	259	477	721	653	905	1195
Dose rate (cGy h^{-1})	115.8	62.9	41.6	73.5	53.0	40.2
Combined dose rate (cGy h^{-1})	156.0	115.9	114.1	—	—	—

To achieve the desired dose of 75 Gy to the mucosa and 55 Gy to the skin, the mould most be worn for a total of 48.0 hours; this gives doses of 74.9 Gy to the mucosa, 55.4 Gy to the centre and 54.8 Gy to the skin—a clinically acceptable compromise.

5.2.5.6 Cylindrical mould. The shaft of a penis of average diameter 3 cm is to be treated over a length of 8 cm at a distance of 1 cm using a coaxial ring cylindrical mould. A surface dose of 50 Gy in 6 days is prescribed, the mould being worn for approximately 15 hours daily.

For this treatment, $d = 3.0$ cm, $D = 5$ cm and $L = 8.0$ cm; thus $L/D = 1.6$ and $d/D = 0.6$. Using equation (5.9) the mg h per 10 Gy at the centre of a ring of diameter of 5 cm is equal to $30 \times 1.08 \times 5^2 =$ 810 mg h. From table 5.3 the multiplying factor is 1.69, thus to deliver a dose of 50 Gy to the surface we need $810 \times 1.69 \times 5 = 6844.5$ mg h. To deliver this dose in 90 hours, 76.05 mg of radium or radium-equivalent sources are required.

Distribution. For a treating distance of 1 cm, the separation between the rings is 2 cm, thus we require four rings, each of the two middle rings containing 20% of the total activity, i.e. 15.2 mg, and the two outer rings each containing 22.9 mg (figure 5.4). The circumference of the mould is $5\pi = 15.7$ cm. Thus we can use 15 type F-1 tubes in each end ring (22.5 mg Ra eq per ring) and 10 type F-1 tubes in each middle ring (15 mg Ra eq per ring)—a total content activity of 75 mg. The actual treatment time, therefore, is increased to 6844.5/75 which equals 91.3 hours and the mould is to be worn for 15 hours 18 mins per day.

The dose to the urethra can be obtained by taking $d = 0$. Therefore, from table 5.3 the factor is 2.0, thus the dose received is $(1.69/2.0) \times 50 = 42.25$ Gy. If the dose throughout the volume is not sufficiently even, then the treatment distance needs to be increased.

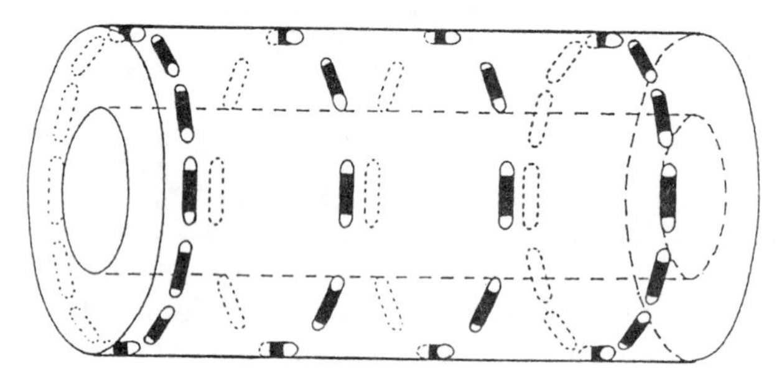

Figure 5.4 Diagram of cylindrical mould. (Reproduced from Meredith (1967) by permission of Churchill Livingstone Ltd, Edinburgh.)

5.3 Other Dosage Systems

The other main dosage system currently used is that due to Quimby (1944). This is simpler in concept than the Manchester system and is based on the dose delivered at various distances below the centre of an applicator when the radium is uniformly distributed over the surface of the applicator. As with the Paterson–Parker dosage system, the Quimby tables were initially tables calculated for various geometry surface applicators in terms of mg h for 1000 R using a specific gamma-ray constant factor of 8.4 R h^{-1} mg^{-1}. These have been subsequently modified using a factor of 1.064 to give the mg h of radium required to give 10 Gy for a filtration of 0.5 mm of platinum (table 5.15). The number of mg h is dependent on the shape of the applicator. With the Quimby tables it is found that less radium is needed to achieve a comparable dose to the Manchester system. However, the variation in dose across the surface can be as much as 45% corner to corner (Glasser *et al* 1961). The following example serves to illustrate the differences between the two systems.

Example 5.2

Calculate the amount of radium required by both systems to treat a lesion of 4×6 cm^2 using a mould 0.5 cm above the surface of the skin. A dose of 60 Gy is required to the surface in six days, the mould being worn

Table 5.15 Quimby dosage system tables for moulds and implants (mg h for 10 Gy at points along a line perpendicular to mould or implant). (Modified from data of Quimby (1944) for radium sources, filtered by 0.5 mm Pt distributed uniformly across the plane.)

Distance from applicator (mm)	Circular applicators (diameter in mm)					
	10	20	30	40	50	60
5	47	80	110	181	234	319
10	145	187	234	319	394	482
15	301	345	426	506	598	725
20	528	577	646	745	846	977
25	782	846	920	1016	1229	1346
30	1160	1224	1298	1404	1522	1665

Distance from applicator (mm)	Square applicators (length of side in mm)					
	10	20	30	40	50	60
5	49	85	122	210	266	372
10	150	200	253	348	431	544
15	314	367	442	544	638	782
20	532	606	686	795	910	1064
25	777	846	952	1075	1213	1458
30	1160	1224	1351	1479	1617	1777

Distance from applicator (mm)	Rectangular applicators (mm × mm)					
	10 × 15	20 × 30	30 × 40	40 × 60	60 × 90	80 × 120
5	54	110	152	305	606	1016
10	157	228	291	453	772	1181
15	317	394	496	664	1005	1442
20	538	628	761	930	1319	1777
25	767	894	1053	1213	1617	2128
30	1181	1266	1420	1617	2054	2660

six hours daily. Clinically it is important to know the dose 0.5 cm below the skin surface.

Consider table 5.16.

Table 5.16 Relevant parameters for Example 5.2.

For a mould of area 24 cm^2	Quimby system	Manchester system
Number of mg h for 10 Gy at 0.5 cm	453	764
Number of mg h for 10 Gy at 1.0 cm	664	1089
Elongation factor	1.000	1.025
Time of application (hours)	36	36
Total number of mg of radium required	75.5	130.5
Depth dose 0.5 cm below surface	68.2	70.2
Dose at 0.5 cm in 36 hours (Gy)	40.9	42.1
Distribution of sources	Uniformly	Three parallel lines

Whilst the Manchester system requires more radium to achieve the dose, the uniformity of dose is better, as is the dose at depth. *It must be emphasised that it is highly inaccurate to arrange the sources in one way and then calculate doses by the tables set out for the other.*

5.4 Beta-emitting Plaques

For many years beta-ray plaques were made using isotopes in the radium series. These were sealed into a metal container in such a way that the wall thickness only allowed beta-particles to emanate from the front surface. The isotopes used were radium D($^{210}_{82}$Pb) and radium E($^{210}_{83}$Bi) which have half-lives of 19.4 years and 5.0 days respectively. Unfortunately, the beta energy was relatively low with a maximum of 1.17 MeV, thus only relatively superficial lesions could be treated. With the availability of artificially produced isotopes, beta-emitting plaques became viable. Friedell *et al* (1951) in the US and Sinclair and Trott (1956) in the UK were some of the early workers who constructed applicators of artificially produced isotopes. Attention was initially concentrated on the use of phosphorus-32 impregnated into plastic sheet as a mould. However, when it proved possible to produce thin foils of strontium-90 bonded in silver this offered a more promising alternative. Applicators using strontium-90 are commercially available (see §2.3.7), both as ophthalmic and surface applicators.

The dosimetry of such beta plaques requires a knowledge of:

(i) the absolute value of dose rate at the surface;
(ii) the variation in dose rate across the surface; and
(iii) the distribution in dose rate in tissue below the surface of the applicator.

In addition, it is important to establish whether bremsstrahlung, gamma-rays or x-rays from impurities in the source contribute appreciably to the dose rate. Two main attempts have been made to derive this knowledge, by theoretical calculations and by measurement. Loevinger *et al* (1956) described in detail methods of calculating the dose distribution around sources, based on the assumption that the point source distribution can be represented by a function of the form

$$J(x) = k\mathrm{e}^{-\nu x}(\nu x)^{-2}$$

where $J(x)$ is the dose in cGy per disintegration at a distance x from a point source of beta-particles and ν is a suitable 'absorption coefficient'. The second term represents the inverse square attenuation. Starting with this assumption, Loevinger developed a series of equations to calculate the dose for a variety of surfaces by summing the contribution to a point P from a series of point sources distributed over a surface S. Thus the dose at P, D_P is given by

$$D_\mathrm{P} = \int_\mathrm{S} A_\mathrm{S} J(x)\mathrm{d}S \quad \mathrm{cGy}$$

where A_S is the surface activity in disintegrations per cm^2 and the integral is extended over the entire surface S (with units of cm^2). This approach, however, has been shown to be only reliable for distances of about half the maximum beta-particle range; for distances greater than this the calculated values are found to be greater than the true values. Detailed study of this topic will be found elsewhere (for example, Loevinger *et al* (1956), which also covers dose calculations due to thin-plane thickness of isotopes and circular discs).

The other approach to determine dose distribution has been by measurement (Friedell *et al* 1951, 1954, Sinclair and Trott 1956, Jones and Dermentzoglou 1971). Due to the rapid fall-off in dose with distance, the measurement of dose rate from a beta-source is difficult and even more complicated in the case of a curved ophthalmic applicator. The method most often adopted is to use an extrapolation chamber, similar in size to the applicator, to make a direct measurement of the surface dose rate of a flat strontium-90 applicator and then, using a suitably constructed beta-scintillation probe, compare the surface dose rate on the flat plaque with that of the ophthalmic applicator. The design of the extrapolation chamber is critical (Loevinger 1953), but if the necessary requirements are met, it is possible to achieve an accuracy of better than ±2%. The scintillation probe can also be used to determine the dose rates at various other points on the surface.

Using this approach, the manufacturers supply ophthalmic applicators with charts which show the variation in dose rate at points over the surface as a percentage of the surface dose at the centre of the applicator. Percentage depth dose measurements are also required for clinical use. These measurements can be made by placing an accurately known thickness of phantom material, say Perspex, in between the applicator and the end of the probe (Jones and Dermentzoglou 1971).

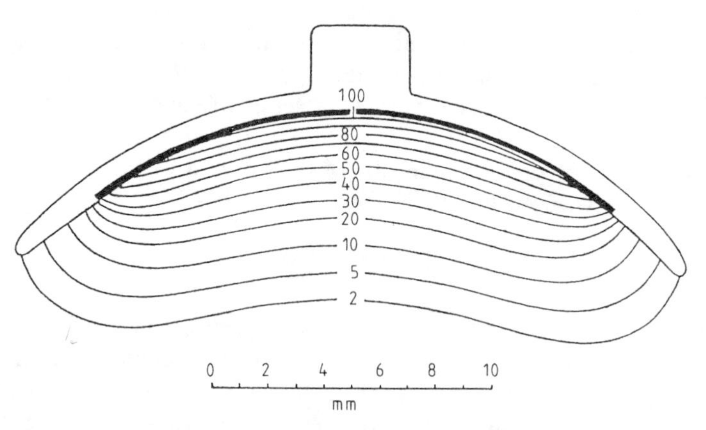

Figure 5.5 Diagrammatic representation of the isodose curves in tissue from a strontium-90 eye applicator. (Reproduced by kind permission of Amersham International plc.)

In order to determine the distribution around a source, film dosimetry is often employed to produce isodensity curves which are then used in conjunction with percentage depth dose data to construct the dose distribution around beta-emitting plaques (figure 5.5). In clinical use care must be taken in applying the applicators since small air gaps or tear fluid can result in major underdosing.

5.5 Gamma-emitting Plaques

Whilst beta-plaques are used in the treatment of such eye conditions as epibular malignant melanamata and corneal vascularisation, the treatment of retinoblastoma requires ophthalmic applicators which give greater penetration. Stallard (1962) developed a technique of suturing cobalt-60 applicators to the eye. Measurements on enucleated eyes showed that on average the depth of the lesion below the retina is approximately 2/3 of the lesion diameter, thus a series of applicators were developed which enabled a treatment of at least 40 Gy in 6 days to be given to the desired depth for each applicator. This depth, the design depth, was calculated by Innes (1962) and each applicator was constructed in such a way that the activity and distribution of the cobalt-60 within the applicator gives the desired dose with as low and as uniform a dose as possible to the sclera.

The dosimetry of these applicators was studied by Casebow (1971) who derived equations which enabled the dose distributions in the eye to be interpolated from calculated arrays of exposures. The calculated distributions were then verified by measuring the exposure distribution using film techniques. The computed values overestimated the exposure at the edges of the applicator due to the effects of increased attenuation in the platinum sheath in this region. In an attempt to correct this overestimate, Casebow introduced a graphically derived correction factor. The dosimetry of cobalt-60 ophthalmic applicators was also studied by Chan *et al* (1972). A theoretical dose distribution was calculated using a Sievert line-source approach, which had the effect of partially correcting for oblique filtration. The results obtained for the central axis depth doses were in close agreement with those found by Innes (1962). These gamma-ray applicators are now being replaced in part by ruthenium-106 beta-ray ophthalmic applicators.

6 Interstitial Therapy

6.1 Introduction

The need for dosimetry in patients treated with implanted radioactive sources became apparent following the experiences of tissue necrosis by early clinical workers. These reactions were soon realised to be due to the inadequate filtration of the beta-rays from radium daughter products. Once these filtration problems had been remedied, the implantation of encapsulated radionuclides became a universally accepted technique.

Much of the early clinical experience was gained at the Radium Institute in Paris by Regaud and colleagues during the early 1920s. They came to the conclusion that effective treatment could be acheived in certain cancers by implanting low-activity radium needles for six to ten days, a dosage regime followed subsequently by many others workers including Paterson and Parker in Manchester.

Clinically, implant techniques are used in the treatment of squamous cell carcinomas of the oral cavity, certain skin and lip tumours, the anus, vaginal wall, urethra, breast and many other sites. Interstitial therapy has many attractions since it provides an effective means of treating tumour tissue to a high dose, whilst at the same time giving a much lower dose of radiation to surrounding tissue. By its very nature, this form of therapy means that dose homogeneity throughout the implant is impossible since there must always be zones of intensely high radiation in the immediate vicinity of each implanted needle or seed. Since these zones usually fall within the target volume they are not clinically significant, but it does mean that a different concept of dose uniformity is required from that adopted for mould techniques. Paterson and Parker in the 1930s considered ways of obtaining a 'uniform dose' within the volume and devised a system of tables and distribution rules analogous to those for surface applicators to achieve this result. At the same time Quimby, who was also studying the problem of interstitial therapy, devised a dosage system which used the minimum dose on the surface of an implant as the basis of dosage calculations. Naturally, since the specification of dose is different between the two systems, the amounts of radium to be used to achieve a given dose differ. Although the two approaches require different amounts of radium to give the same nominal dose, each system is self-consistent and both have been used for many years to calculate the dose absorbed within tissue. For a discussion on the relative merits of each system, the reader should refer to the following publications: Quimby and Castro (1953), Bloedorn (1956), Stovall and Shalek (1968) and Laughlin *et al* (1963).

Initially, these systems were devised for use with radium and radon. With the availability of artificially produced radionuclides in the late 1950s, the use of the tables was extended to other radio-isotopes, such as cobalt-60 (Fletcher *et al* 1954) and gold-198 grains (Shalek *et al* 1957). Further developments in interstitial dosimetry came in the 1960s with the introduction of computers for dose calculations and the availability of such isotopes as iridium-192 for brachytherapy. This led to a different approach being adopted to dose calculation, since computers made it practical to create dose distribution in any plane within the implant, thereby providing the radiotherapist with more information about dose variation within the tumour volume (Shalek and Stovall 1961). Whilst these methods of interstitial dosimetry continue to increase, other dosage systems have been suggested over the past 25 years as alternatives to the Manchester and Quimby systems for use with the artificially produced isotopes such as iridium-192 (Pierquin and Dutreix 1967), iodine-125 (Henschke and Cevc 1968) and gold-198 (Dale 1976).

The popularity of interstitial therapy as a treatment modality has fluctuated over the years. During the 1950s there was a decline in use due to concern about the radiation exposure staff were receiving in these procedures. However, since that time developments have taken place in implantation techniques which make it possible to reduce exposure to staff. In these techniques the emphasis has been on implanting inactive rigid needles or flexible tubes into the tumours which can then be loaded with a suitable radioactive source at a later time. The use of such after-loading techniques in interstitial work was pioneered by Henschke in the United States and Pierquin and colleagues in Paris—although it is of historic interest to note that in 1910 Abbe reported on the insertion of radium, contained in glass capsules, into celluloid tubes previously inserted into a tumour. Various workers over the last 40 years made numerous attempts to devise other after-loading systems (for example, Hames (1937), Morton *et al* (1951), Morphis (1960), Suit *et al* (1961), Allt and Hunt (1963) and Bier *et al* (1973)). Such an increase in these techniques means that implant techniques are now being used more, although it must be emphasised that not all clinical sites lend themselves to after-loading techniques.

6.2 Temporary Needle Implants Using Radium or Radium Substitutes

6.2.1 Principles of the Manchester system and dosage tables

The dosage system devised by Paterson and Parker (1938) for interstitial therapy was a natural extension of their work on surface applicators. In this situation, the definition of 'stated dose' is not the same as for applicators; it is now defined to be 10% above the minimum dose in the plane of calculation and represents the effective minimum dose throughout the slab of tissue. It is, however, well recognised that the maximum dose rate in the plane of the sources will be of the order of 30% higher than the stated dose and that the dose to points close to the needles will be five times higher than the stated dose. This work, again based on radium, involved calculating the exposure rate in air expressed in roentgens per unit time and made the assumption that differences in tissue dose due to photon attenuation and the build-up factor could be neglected. In addition, no attempt was made to make allowance for any beta-radiation which may penetrate the wall of the source. The devised system is based on two separate approaches, namely for planar and volume implants.

6.2.1.1 Planar implants. For interstitial therapy Paterson and Parker considered the 'uniformity' of dose achieved in a plane 5 mm from, and parallel to, the plane of the implant. The sources are therefore in the central plane of a 1 cm thick slab of tissue and the plane of calculation is bounded by the projection of the peripheral needles onto that plane. The dosage tables derived give the number of mg h of radium required in this central plane to produce a dose of 1000 R in the plane of calculation, for implants of different areas, the area being defined by the width of the implant and the distance between the crossing needles. These tables have been revised to give the mg h required to yield a dose of 10 Gy (Gibbs and Massey 1980), and recently into tables of air kerma at 1 m per unit system absorbed dose in units of μGy Gy^{-1} when the source strength is specified in air kerma at 1 m (Massey *et al* 1985), as detailed in §5.2.1. Due to the similarity in approach to deriving uniformity of dose between surface applicators and planar implants, the dosage tables for single planar implants are numerically identical to the applicator tables, for 5 mm treating distance; thus table 5.1 is applicable.

In many interstitial techniques, it is not possible to encompass the lesion with a completed ring or rectangle of sources, thus allowance has to be made for the fact that the area in the plane of calculation receiving the stated dose is smaller than the geometric area defined by the needles. To overcome this problem Paterson and Parker recommended that when using the tables of the Manchester system the geometric area be reduced by 10% for each uncrossed end. For implants with one end uncrossed the geometric area is defined as the width multiplied by the distance from the crossing needle to the distant active end; whereas for both ends uncrossed the length is the active length of the needle.

To treat thicker lesions, i.e. of thickness greater than 10 mm, consideration has to be given to factors which affect the dose and its uniformity in the plane of calculation, which is considered to be equidistant from the two single planar implants. By studying the ratio of the maximum to minimum dose for different areas and separations between the two planes of the implant, Parker found that for separations not exceeding 15 mm the two-plane technique provides reasonable homogeneity for all areas (Meredith 1947). Here the maximum dose is defined as the highest dose volume found in the plane of needles but 5 mm distant from the nearest needle. This gives a value 30–40% higher than the average exposure rate at 5 mm from the plane, whilst the minimum is in the plane midway between the sources. For separations greater than 1.5 cm, the dose in the mid-plane means that variations of more than 40% are found, which may or may not be clinically acceptable since the dose at the mid-plane is lower by the percentages shown in table 6.1.

Table 6.1 Separation factors.

Separation between planes (cm)	Area (cm^2)	Percentage reduction of dose in mid-plane	Separation factor
1.5	0–50	10	1.25
2.0	0–25	20	1.40
	25–50	10	
2.5	0–25	30	1.50
	25–80	20	

If the lesion thickness requires a two-plane implant, then for a separation of 1 cm between the source planes the total amount of radionuclide required, in mg h per 10 Gy, will be that given in the tables, the sources being divided equally between the two planes. If, however, the separation is greater than 1 cm, then the amount of radionuclide required to give a certain dose to the mid-plane has to be increased by the separation factor. An analysis of over 100 cases of implants of various areas revealed that it was possible to assess the dose delivered in a two-plane implant by using the single-plane data and a conversion factor. The result showed that the value of the conversion factor varied little with area for a given separation and thus it was possible to use a single conversion factor for each separation.

The physical basis of the distribution of the sources within the implant indicated that the arrangement of the sources should have the crossing needles at the level of the ends of the active length of the needles and should not be more than 1 cm away from the active ends. However, recent work by Doss and Richman (1979) has shown the position of this cross needle is critical and that variations of the order of 2–3 mm are crucial. They recommend that the best coverage is achieved if the crossing needle is either directly at or slightly beyond the physical ends of the needles.

6.2.1.2 Volume implants. When the tumour volume has a thickness greater than 2.5 cm, then it is impossible to achieve an acceptable degree of dose homogeneity using planar implants. Consequently, the method of considering dose based on surface applicator ideas is inappropriate. In 1931 Souttar published a mathematical approach for determining the amount of radium required and how it should be distributed in the volume by considering the dose variation within a sphere of radioactive fluid. To achieve a uniform dose rate throughout the volume he suggested that the volume should be separated into two regions; an inner core and an outer rind, containing radioactive fluids of differing concentrations. Initially Souttar suggested that a suitable arrangement for a sphere of 4 cm radius would be to have an inner core of radius 3 cm filled with a radioactive fluid of concentration ρ mg cm^{-3}, with the spherical shell between the radii 3 and 4 cm (the rind) at a density of 2.5ρ mg cm^{-3}.

Parker used this method to establish the distribution rules for volume implants but chose a different thickness of shell from that used by Souttar and increased the concentration of radioactive fluid in the outer shell to 3ρ mg cm^{-3}. To determine the thickness of shell (A) required to treat a spherical volume of radius 5 cm with a homogeneous dose, the total volume can be considered to consist of an inner spherical core (B) of radius R cm containing a fluid of concentration ρ mg cm^{-3} and a shell of thickness $(5 - R)$ cm of concentration 3ρ mg cm^{-3}. The radial dosage variation, as a percentage of the dose at the centre of the sphere for the 5 cm sphere can be obtained by using equation (3.12) in §3.2.5, whilst the dose variation due to the small sphere, $(5 - R)$ cm diameter, can be obtained by a similar method. The difference between the two results defines the dose distribution due to the shell. By summing the dose due to the smaller sphere and the shell, the radial dosage determined using the Souttar approach can be obtained as given in table 6.2 for $R = 5$ cm. The dose is also found to be dependent on the choice of R. This approach led Parker to conclude that the optimum thickness of the shell should be 1/10 of the diameter of the volume. It is also noted that the effective coverage of the volume is 3 to 4 mm less than the actual volume of the sphere.

Table 6.2 Dose uniformity within a sphere of radiation of radius 5 cm.

Type of volume for which dose is calculated	Dose (cGy h^{-1}) Distance from centre of sphere (cm)								
	0.0	1.0	2.0	3.0	4.0	5.0	6.0	8.0	10.0
Sphere A, $r = 5$ cm	20.00	19.73	18.90	17.39	14.94	10.00	5.60	2.85	1.76
Sphere B, $r = 4$ cm	16.00	15.66	14.59	12.54	8.00	4.04	2.64	1.41	0.88
Shell (A – B)	4.00	4.07	4.31	4.85	6.94	5.96	2.96	1.44	0.88
4 cm sphere radiation dose + 3 times shell dose	28.00	27.87	27.52	27.09	28.82	21.92	11.52	5.73	3.52

A similar approach was adopted by Parker to determine the uniformity of dose within a cylindrical volume. By integrating equation (3.9), as indicated in §3.2.4, the dose variation can be obtained by subtracting the contribution due to a smaller volume from that due to a larger volume, thereby determining the relative dose due to the core, rind and also the ends of the cylinder. This general approach has been used to obtain a uniform dose distribution throughout various shaped volumes. A more detailed approach to the dosimetry of various shaped volumes can be found in the article by Meredith (1947).

These various studies showed that the mean dose rate at the geometric centre within solids of the same volume and radioactive content is independent of shape, except for a small correction factor determined by the elongation of the volume. Parker therefore considered that to a first approximation it was sufficiently accurate to regard the value at the centre of the volume as the average dose throughout the volume, and that the dose rate to any solid (treated on the rind and core principle), in mg h for 1000 R, can be given by

$$\text{mg h per 1000 R} = 21.57\ V^{2/3}\ \mathrm{e}^{0.07(E-1)} \qquad (6.1)$$

where V is the volume in cm^3 and E is the elongation factor which is equal to the ratio of the longest to the shortest principal axis.

The whole of this approach is based on a volume filled with a radioactive fluid. When the transition is made to using arrangements of discrete sources, areas of high local dose are bound to occur. We can therefore no longer consider the average dose throughout the volume but rather the minimum dose in the volume, which in the Manchester system for volume implants is taken to be 10% higher than the absolute minimum in the effective volume. Again, by considering spherical and cylindrical volumes, it was found that if lines are constructed radiating from the centre of the volume, and if such a line passed between sources, then the minimum dose at the surface of the volume is 10% below the stated dose, and that the maximum on such a line is 10% above the stated dose, although it could be as much as 20% above near the centre of the implant. This study also showed that the mean net minimum dose rate was 78% of the average dose rate when compared with the corresponding fluid distribution, thus the dose equation (6.1) must be altered accordingly so that it refers to the minimum dose.

During the same study it was also noted that the linear dimensions of the implant were smaller than the basic volume by a factor of 0.9. This factor also has to be incorporated into the dose equation. These corrections, together with the overall factor of 1.08 as discussed in §5.2, mean that the number of mg h required to give an absorbed dose of 10 Gy is given by

$$\begin{aligned} \text{mg h per 10 Gy} &= 21.57(100/78)(10/9)^2 V^{2/3}\mathrm{e}^{0.07(E-1)} \times 1.08 \\ &= 36.9\ V^{2/3}\mathrm{e}^{0.07(E-1)}. \end{aligned} \qquad (6.2)$$

For volume implants, the number of mg h for 10 Gy are given in table 6.3(a), together with the correction factors to account for elongation (table 6.3(b)).

When volume implants are used it is also assumed that it is possible to use crossing needles at both ends of the implant. This is not always possible and so corrections need to be made to account for the dimensions of the volume being irradiated. Meredith and Stephenson (1947) studied this, together with the positioning of the crossing needles, and found that the volume should be reduced by 7.5% for each open end and that ideally the needles should be inserted at the level of the active end of the rind and core needles. If, however, this was not practical and they have to be placed at the tips of the needles, then the crossing strength should be increased.

Table 6.3(*a*) Number of mg h for 10 Gy for volume implants.

Volume (cm^3)	Number of mg h per 10 Gy	Volume (cm^3)	Number of mg h per 10 Gy
1	36.9	60	565
3	76.7	80	685
5	108	100	794
10	171	140	994
15	224	180	1175
20	272	220	1344
30	356	260	1502
40	432	300	1652
50	500		

Table 6.3(*b*) Correction factors to account for elongation.

Ratio of longest to shortest axis	1.25	1.50	2.0	2.5	3.0
Elongation factor, E	1.02	1.03	1.06	1.10	1.15

6.2.2 The Paterson–Parker distribution rules

The dosage tables mentioned above only apply when used in conjunction with the following distribution rules.

6.2.2.1 Rules for planar implants.

(i) Sources should be placed in a single plane in such a way that a certain proportion of the total amount of radionuclide is around the periphery, whilst the remainder is placed as evenly as possible. The relative proportions are given in table 6.4

Table 6.4 Distribution of sources for planar implants.

Area	Relative proportion of radionuclide Periphery	Centre
Area < 25 cm^2	2/3	1/3
25 cm^2 $<$ Area $<$ 100 cm^2	1/2	1/2
Area $>$ 100 cm^2	1/3	2/3

(ii) Needles should be in parallel rows at a spacing not greater than 1 cm, i.e. $2h$, where h is the effective treating distance of 0.5 cm.
(iii) Crossing needles should be across the active ends ideally, but not more than 1 cm from them.
(iv) If the ends of the implant cannot be 'crossed', 10% must be deducted from the area for each uncrossed end and the reduced area used when reading the tables.
(v) For implants using small sources (seeds or short needles) the distance between active ends should not exceed 1 cm.
(vi) If two planes are used, the separate planes should be arranged as above, parallel to each other, and if they differ in area then the area used to obtain the amount of activity required is the average of the two and the activity is divided pro rata to each area. To account for the separation between the planes a separation factor (table 6.1) must be used to correct the mg h for 10 Gy.

6.2.2.2 Rules for volume implants. The volume is considered to be formed from a rind which comprises the entire surface and a central core. In the case of a sphere the rind is the surface, whereas with a cylinder it consists of a 'belt', the curved surface of the cylinder and two flat ends. The total activity required is divided into a number of equal parts and then distributed throughout the volume in accordance with table 6.5.

In the case of a cuboid, the amount of radionuclide required should be divided into eight parts—one part for each side and each end and two parts for the core. Since a cuboid or rectangular block of tissue can be considered as a multiplane implant, rather than a rind and core arrangement, the mg h of radium or radium-equivalent per 10 Gy determined from the volume tables should be subdivided into three parts for each outer plane and two parts for each inner plane.

Having allocated the radionuclide to the various regions of the volume, the sources are distributed according to the following rules.

Table 6.5 Distribution of sources for volume implants.

Volume shape	Distribution of sources (parts)			
Sphere		Rind 6		Core 2
Cylinder	Belt	Core	End A	End B
Crossed at active ends	4	2	1	1
1 open end	4	2	1	0
2 open ends	4	2	0	0
Crossed at needle tips	4	2	2	2
1 open end	4	2	2	0

(i) The sources on each face must be spaced as evenly as possible.

(ii) The sources must be distributed throughout the core and *not* just at the centre.

(iii) The sources on each surface must be spaced evenly and not more than 1.0 to 1.5 cm apart.

(iv) For a cylindrical implant, there must be *not less* than eight needles in the belt and four in the core.

(v) For cuboid implants arranged as a multiplane implant, the distribution of the sources is as for planar implants.

(vi) Where it is impracticable to close one or both ends of a cylindrical volume, the geometric volume of the implant (i.e. the cross-sectional area multiplied by the active length of the belt and core needles) must be reduced for dosage computation purposes by 7.5% for each open end. In this situation the sources are divided into parts as shown in table 6.5.

(vii) In certain clinical applications, where it is impossible to implant sources into the core but practical to use an intracavitary applicator as the core, then it is possible to place the core sources at the centre of the volume, provided it is enclosed in a container whose diameter is equal to at least half the diameter of the cylinder. This could be useful in the treatment of an epithelial tumour of the anus.

Oddie (1940) has extended this dosage system to irradiate cones, prisms, pyramids and tetrahedra of tissue. This work showed that it is possible to use the same volume dosage graph, provided a certain correction is applied dependent on the size of the volumes.

6.2.2.3 Influence of source linear activity. To achieve the Manchester recommended dose of 65 Gy (7000 R) in six to eight days for interstitial therapy, the needles should ideally have linear activities of 0.66 and 0.33 mg cm^{-1}. In the United States radium needles were manufactured with linear activities of 1.0 and 0.5 mg cm^{-1}. The use of such sources was studied by Quimby and Castro (1953) who carried out a systematic investigation of dosage distribution in planar and volume implants for radium distributions approximating the Manchester system, but only using needles having a linear activity of 1 mg cm^{-1}. In this work, three distributions were studied for single-and two-plane implants by the following methods.

(i) The first used essentially the same division of radium between the periphery and the central area as the Manchester system, but with the spacing between needles varying between 1.2 and 1.7 cm instead of the 1.0 cm or less recommended.
(ii) Alternatively, the recommended spacing of 1.0 cm was kept, which resulted in the ratio of the central to peripheral dose being considerably higher than the rules specify.
(iii) Lastly, the spacing was kept at 1 cm, but using half-strength needles for the centre which, for smaller fields, resulted in the peripheral dose being, on average, three times as much as the central.

The results of this study showed that none of the plans were quite as good as the true Paterson and Parker arrangements. However, the variations seen were clinically acceptable, although the increased linear activity of the needles resulted in higher dose rates which necessitated giving a dose of between 5000 and 6000 R in three to four days to produce the same biological effect. Similar variations were noted when two-plane and volume implants were considered.

Howells and Oliver (1964) also studied the dosimetric implications of single-plane implants, using needles of the same uniform linear activity, and indicated that the rules are approximately fulfilled if a spacing of 1.5 cm between the needles is used rather than 1.0 cm, although the dose in the treatment plane then varied by +15% to −20% from the average value, instead of the ±10%. It has, however, been suggested that since tumours in practice are not of uniform 1 cm thickness but tend to be lens-shaped, a clinically accepted treatment can be achieved using the same linear activity sources spaced 1.0 cm apart. This arrangement would then create a lens-shape dose distribution in keeping with clinical requirements. To achieve this result, Howells and Oliver developed rules to provide values of the dimensions of this isodose surface for an average dose rate calculated from the Paterson and Parker area tables and the actual radium activity used.

The effect of needle spacing, whilst maintaining differences in the linear activity of the needles, has also been studied by Richman *et al* (1980). In this work, a spacing of 0.8 and 0.7 cm between parallel needles was used and it was found that this increased the area covered by 85% of maximum dose rate by 14% to 18%.

6.2.3 Other radium systems

The other radium dosage system which has been used extensively for interstitial therapy is that due to Quimby (1944). In this system sources of equal intensity are distributed evenly within the area to be treated for planar implants, and evenly throughout the volume for volume implants. In the case of planar implants, the needles should be spaced parallel to each other, about 1.5 cm apart. They should be in a cylindrical configuration for volume implants and should extend throughout the volume, the outer ones being at or outside the margins of lesion. This approach has certain advantages over the Paterson and Parker system for very small volumes, since it is often impracticable to implant at least eight needles in the lesion. As volumes become progressively larger, the uniform distribution approaches the Manchester distribution. However, this sort of distribution produces a relatively greater central dose than the Paterson–Parker method.

A dosage table, based on the assumption that dose is proportional to the square-root of the volume, was initially devised by Quimby to give the number of mg h of radium required to give 1000 R, using the appropriate elongation factor where necessary. This was subsequently corrected to give the mg h for 10 Gy (table 6.6). By comparison with Paterson–Parker tables it will be noted that for a given dose specification, a larger number of mg of radium will be required to give the same dose with the Quimby system as with the Manchester system. These differences are partly accounted for by the difference in source distribution and partly by the position of dose specification. Unlike the Manchester system, the Quimby dose system defines the 'stated dose' to be the *minimum* dose which occurs in the actual implanted region. If one considers, for example, a volume of less than 10 cm^3 satisfying the distribution rules of each system, the stated exposure in the Quimby system may be as much as 50% greater than in the Paterson–Parker system. This difference is best illustrated by considering a practical example.

Table 6.6 Number of mg h required for *minimum* absorbed dose of 10 Gy in a volume implant. (Modified from data of Quimby (1944).)

Volume (cm^3)	Number of mg h per 10 Gy	Volume (cm^3)	Number of mg h per 10 Gy
1	70	60	798
3	155	80	926
5	213	100	1064
10	340	140	1290
15	415	180	1500
20	468	220	1675
30	575	260	1825
40	660	300	1915
50	738		

Example 6.1

A cylindrical volume 4 cm in diameter and 5 cm long is to be treated to a dose of 60 Gy in 160 hours, one end being uncrossed. Table 6.7 gives the relevant parameters in both systems.

Table 6.7 Parameters for Example 6.1.

Parameter	Manchester	Quimby
Volume (cm^3)	62.8	62.8
Uncrossed end factor	1.075	1.000
Effective volume (cm^3)	58.4†	62.8
Number of mg h for 10 Gy	555.6	845
Elongation factor	1.02	1.02
Number of mg for 60 Gy in 160 hours	21.3	32.3

† Volume treated reduced by 7.5% for the uncrossed end in this system.

The results of this example contrast with the requirements for a planar implant as shown in Example 5.2 where the number of mg of radium or radium-equivalent required are less when the dose is calculated by the

Quimby system rather than the Manchester system. *It is therefore imperative that one system is used to the exclusion of the other within one centre.*

6.2.4 Radium substitutes

Whilst the interstitial dosage tables were primarily developed for use with radium and radon seeds, they can also be used with other radionuclides. Myers (1948) discussed the use of radioactive needles containing cobalt-60, whilst Fletcher *et al* (1954) highlighted some of the problems encountered in using cobalt-60 needles for interstitial and intracavitary work. Due to the low beta-ray energy emitted by cobalt-60, it was possible to construct thinner needles which were initially thought to be advantageous. In order to use the Paterson–Parker tables a conversion factor, the ratio of the specified gamma-ray constants of radium to those of unfiltered cobalt-60, was used to determine the activity of cobalt-60 required to achieve a comparable dose.

Another factor considered was the difference in the build-up factor between radium-226 and cobalt-60, a variation later reported by Hale (1958) and others. From the measurements made it was found that the Paterson–Parker system could be used, since the variations were well within the accepted limits of the dosage system. It was recognised, however, that great care had to be exercised in replacing well established radium techniques, since the smaller diameter needles tended to produce more acute radiation slough around the needles due to their higher surface doses. This led to the redesign of the original needles to make them of greater diameter, i.e. a change from 16 to 14 gauge steel. Needles currently used for interstitial therapy contain caesium-137. By defining their activity in terms of mg Ra eq, the dosage tables of Paterson–Parker can be used providing account is taken of the differences in exposure/air kerma rate as discussed in §5.2.4. Replacements for radon seeds for permanent implants have been achieved by using gold-198 (see §6.4). The Paterson–Parker tables can also be used with other medium half-life isotopes, such as iridium-192 (Meurk *et al* 1957), but with these isotopes the dose tables have to be revised to give the millicurie–hours accumulated with time per initial mg Ra eq of the isotope.

6.2.5 Clinical examples

To illustrate clinical examples of implant dosimetry using the Manchester system use will be made of currently available CDCS caesium needles marketed by Amersham International plc. This range of sources (as detailed in table 6.8) provides for activity of 3.7 MBq mm^{-1} (1 mCi cm^{-1}), which is approximately equivalent to 0.33 mg Ra eq cm^{-1}, and 7.4 MBq mm^{-1} (2 mCi cm^{-1}) equivalent to 0.66 mg Ra eq cm^{-1}. In addition to caesium needles, the range includes differentially loaded sources which have been designed for use in areas where it is impossible to cross the ends in an implant. One such design of needle, the Indian club needle, divides the active length into three or four cells and loads the end cell with a linear activity three times that of the others. The net effect of this arrangement is to create a distribution which has the shape of an Indian club. The use of these needles in an implant means that no allowance needs to be made when it is clinically impossible to cross one

Table 6.8 Characteristics of caesium-137 CDCS needle sources (external diameter 1.85 mm, stainless steel wall thickness 0.6 mm). (Manufactured by Amersham International plc.)

CDCS type	Linear activity (MBq mm^{-1}; mCi mm^{-1} figures in brackets)	Content activity (MBq)	Air kerma rate at 1 m ($\mu Gy\,h^{-1}$)	Exposure rate at 1 m ($mR\,h^{-1}$)	mg Ra eq	Total length (mm)	Actual length (mm)
Linearly loaded needles							
A-1		55.5	3.5	0.4	0.5	24.5	15.0
A-4	3.7	83.3	5.2	0.6	0.75	32.0	22.5
A-2	(0.1)	111.0	7.0	0.8	1.0	40.0	30.0
A-3		166.5	10.5	1.2	1.5	55.0	45.0
B-1		111.0	7.0	0.8	1.0	24.5	15.0
B-4	7.4	166.5	10.5	1.2	1.5	32.0	22.5
B-2	(0.2)	222.0	14.0	1.6	2.0	40.0	30.0
B-3		333.0	21.8	2.5	3.0	55.0	45.0
Differentially loaded needles							
D-1	Dumb-bell	166.5	10.5	1.2	1.5	40.0	30.0
D-2		222.0	14.0	1.6	2.0	55.0	45.0
D-3	Indian club	166.5	10.5	1.2	1.5	45.0	35.0
D-4		222.0	14.0	1.6	2.0	60.0	50.0

Notes Cell arrangements for type D needles:

Type D-1 Cell B + Cell A + Cell B
Type D-2 Cell B + Cell A + Cell A + Cell B
Type D-3 Cell A + Cell A + Cell C
Type D-4 Cell A + Cell A + Cell A + Cell C.

Cell contents:

Cell A 55.5 MBq (1.5 mCi), active length 15 mm
Cell B 55.5 MBq (1.5 mCi), active length 7.5 mm
Cell C 55.5 MBq (1.5 mCi), active length 5.0 mm.

end, as is the case when implanting a lesion of the tongue. The other differentially loaded sources, 'dumb-bell' sources, are those in which each of the end cells has a linear activity of twice the linear activity of the central cell or cells of the needle. These sources produce a dose distribution in the shape of a dumb-bell. In devising the treatment the ideal distribution is planned prior to the insertion, but to calculate the actual time of the treatment radiographs are necessary to determine the actual volume implanted.

6.2.5.1 Single-plane implant. A tumour of the tongue of area 3 cm wide by 5 cm long is to be treated with a single-plane implant to a prescribed dose of 60 Gy in about seven days. Clinically it is not possible to cross the lower end of the implant. Calculate the treatment time using the sources available to carry out this treatment.

(i) *Planned treatment.* The area is to be implanted with four needles (since $2h = 1$ cm) and one crossing needle. The needles to be used are the A-3 and B-3 type since they treat a length of 5 cm if the crossing needle is at the tip together with a B-2 type crossing needle.

Area of implant = 15 cm^2, effective area is 15 cm^2 less 10% for one uncrossed end = 13.5 cm^2. From table 5.1, mg Ra eq h for 10 Gy = 303.5. Therefore the total mg h required = 303.5×6. Total mg Ra eq required for treatment in seven days = $303.5 \times 6/168 = 10.8$ mg Ra eq.

(ii) *Source distribution.* Since the area is less than 25 cm^2, we know from table 6.4 that 2/3 of the activity, 7.2 mg, is to be placed on the periphery and the remaining 3.6 mg at the centre. Thus the sources used are: on periphery, two B-3 type needles plus one B-2 type needle, 8 mg; in centre, two A-3 type needles, 3.0 mg.

Since the sources used have a wall thickness of 0.6 mm, the effective mg Ra eq used is $11.0 \times 0.98 = 10.8$ mg. This is the required number of mg Ra eq to give the desired treatment in seven days.

(iii) *Actual treatment.* Radiographs of the implant show that the average width of the implant was 3.1 cm and that the crossing needle has been implanted across the active ends. Thus: the actual area implanted $= 4.5(3.4 + 2.8)/2 = 13.95$ cm^2; corrected area for an uncrossed end = 13.95 − 10% = 12.6 cm^2; source strength required in mg Ra eq for 60 Gy = 6×290. Therefore, actual treatment time = $6 \times 290/10.8 =$ 161 hours, i.e. 6 days 17 hours.

6.2.5.2 Two-plane implant. A two-plane implant is used to treat an anal tumour of area 2.5×3.5 cm^2 and 2.5 cm thick. The area of the implant is 4×4.5 cm^2 and both ends are uncrossed. Calculate the actual duration of insertion to achieve a minimum dose of 60 Gy to the treated volume in approximately six days.

(i) *Planned treatment.* Area of implant = 18 cm^2, effective area is 18.0 cm^2 less 20% = 14.4 cm^2. From table 5.1 the number of mg h for 10 Gy = 317. Since the separation between planes is 2 cm, the separation factor = 1.4, thus the required number of mg Ra eq for the treatment is $317 \times 6 \times 1.4/144 = 18.5$ mg Ra eq.

The activity is to be equally divided between the two planes, thus each plane should ideally contain 9.25 mg, with 2/3 of this activity on the periphery and the remainder in three rows spaced at a distance of 1 cm. The needles chosen, therefore, are two type B-3 needles for the periphery and three type A-3 needles for the rows. Since these needles have a wall thickness of 0.6 mm, the total activity in each plane is $(2 \times 3 + 3 \times 1.5) \times 0.98 = 10.3$ mg Ra eq. Thus the final treatment time is $317 \times 6 \times 1.4/20.6 = 129.2$ hours, i.e. 5 days 9 hours 15 min.

No allowance is made in this example for dose rate, as suggested by Paterson, and it is assumed that the actual implant was as for the plan.

6.2.5.3 Volume implant. A tongue lesion is to be treated to a dose of 60 Gy in about six days using a cylindrical volume implant with the lower end uncrossed. The surface area of the induration is approximately 3.5 cm in diameter and the lesion is 3.0 cm thick. Calculate the treatment time to achieve this dose.

(i) *Planned treatment.* Volume = $\pi \times 3.0(3.5/2)^2 = 28.9$ cm^3, effective volume is 28.9 cm^3 less 7.5% = 26.7 cm^3. Using equation (6.2) or by interpolating in table 6.3(a), the number of mg Ra eq h for 10 Gy = 330. The total number of mg Ra eq required for the prescribed treatment = $330 \times 6/144 = 13.75$ mg Ra eq.

(ii) *Distribution.* The belt and core of this implant will be implanted with type A-2 sources and the crossing needles will be at the tip of the needles. From table 6.5 the sources are to be distributed so that there are four parts in the belt, two parts in the core and two parts at the end. Since the belt must contain not less than eight needles and the core four sources, the distribution is as follows: eight type A-2 sources in the belt = 8 mg; four type A-2 sources in the core = 4 mg; and four type A-2 sources at the end = 4 mg. Total number of mg = 16.

(iii) *Actual treatment.* The actual implant size measured on the films was 3.4×3.8 cm^2 with the needles crossed at the tips. Since the tip of the needle to the distant active end is 3.5 cm, then the volume implanted is $n \times 3.4 \times 3.8 \times 3.5/4 = 35.5$ cm^3. The effective volume of the implant is 36.5 cm^3 less 7.5% = 32.9 cm^3. For this volume the number of mg h of radium-equivalent for 60 Gy = 2271. Since the total activity inserted was 16 mg Ra eq, the time for the duration of the implant is 2271/16 = 141.9 hours, i.e. 5 days 22 hours.

6.3 Temporary Implants Using Iridium Wire, Hair-pins and Seeds

6.3.1 Clinical techniques

The use of radioactive needles in interstitial therapy has decreased in recent years due in part to the problems associated with the irradiation of medical and nursing personnel during the implantation procedure. An increasing proportion of interstitial brachytherapy is now performed using after-loading techniques with such isotopes as artificially produced iridium-192. This work has been pioneered over the past 20 years by Pierquin and colleagues in Paris and Henschke in New York who have developed techniques which enable the radioactive sources to be loaded into the lesion, either at the end of the operating procedure or once the patient has returned to the ward. Such techniques have led to improved radiation protection for the majority of staff without detriment to patient treatment, although as shown in table 7.7, §7.4, manual after-loading techniques do not effectively reduce the exposure to ward nursing staff. To improve this situation remote after-loading systems have been developed for implant techniques, for example the Micro Selectron manufactured by Nucletron.

Depending on the location and size of the tumour, three main implant techniques are used. In these techniques, the initial stage is to implant a non-radioactive applicator which is subsequently loaded with iridium-192 in the form of wires, hair-pins and seeds contained within nylon ribbon.

6.3.1.1 Guide gutter technique. This technique was developed by Pierquin and Chassagne and is typically used in the implantation of such lesions as carcinoma of the tongue. The guides consist of incomplete cylinders of stainless steel (figure 6.1) which form channels into which the iridium, in the form of 0.6 mm thick hair-pins or single pins, is inserted once the radiotherapist is satisfied radiologically that the guides are in the correct position. Once inserted, the iridium hair-pins or single pins are sutured into position and the guides withdrawn without disturbing the location of the radioactive wires.

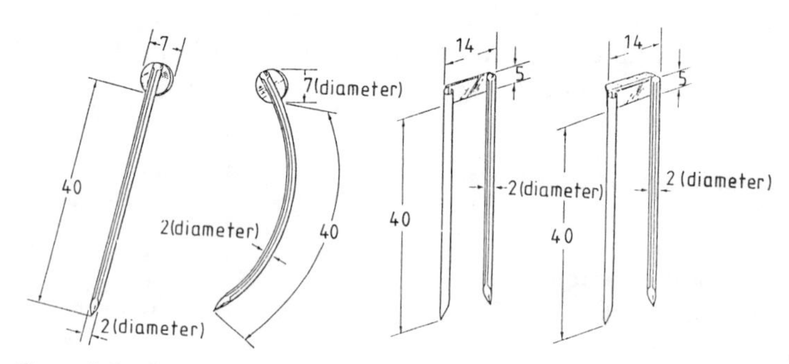

Figure 6.1 Slotted steel guides for use with iridium-192 pins (dimensions in mm). (Reproduced by kind permission of Amersham International plc.)

6.3.1.2 Plastic tube technique. Initially perfected by Henschke *et al* (1963), this technique consists of placing plastic tubes into the patient into which thin 0.3 cm diameter wire is later inserted, the wire first being encapsulated within a narrow-bore plastic tube. The first step in the

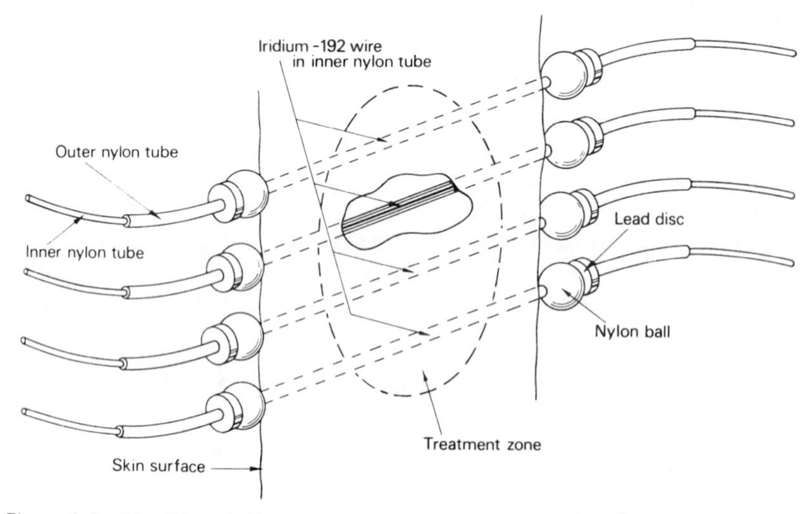

Figure 6.2 The Pierquin/Paine after-loading technique—typical assembly. (Reproduced by kind permission of Amersham International plc.)

procedure, described by Pierquin *et al* (1971) and Paine (1972), consists of inserting a steel guide needle into the tumour. The 1.6 mm diameter outer plastic tubing is placed on a steel mandrel in the guide needle and pulled through the tumour as the guide needle is removed, thereby leaving the outer plastic tube in position in the tumour. The treatment volume is fully implanted using this method and the implant is radiographically checked by temporarily loading each tube with a non-radioactive marker wire to ensure a satisfactory arrangement. Appropriate lengths of iridium-192 wire are sealed into narrower plastic tube of diameter 0.85 mm, using a loading device such as that described by Darby *et al* (1972), and then loaded into the outer plastic tube. The ends of the outer tubes are then clamped with lead discs which are separated from the skin by spherical nylon beads (figure 6.2).

6.3.1.3 Hyperdermic needle technique. In certain sites, such as tumours of the mucous membranes, the use of plastic tubing has certain disadvantages. To overcome these Pierquin developed a technique involving the use of nickel applicators bevelled at both ends. These applicators have an outer diameter of 0.8 mm and internal diameter 0.5 mm, into which 0.3 mm iridium-192 wires are inserted. Since these are rigid applicators which have both entrance and exit points in the skin, the radiotherapist can obtain the desired spacing and ensure the applicators are parallel to each other visually. When the optimum geometry has been achieved, one end of the applicator is closed with a lead cap and the other is used for the insertion of the iridium wire which is then similarly sealed.

The use of stainless steel needles which are after-loaded with iridium-192 wire was originally suggested by Henschke and has been adapted by numerous workers (including Delclos (1982, 1984)) for the treatment of tumours which can only be approached from one side, such as tumours of the vagina, anal margin and nasal septum. In this technique an empty stainless steel needle with a Teflon or nylon button attached is inserted into the tumour and the needle secured in place by suturing it to the skin using the hole in the button. The needle is crimped at one end to position the iridium-192 insert and to prevent it from dropping out (figure 6.3). The iridium wire, contained within a plastic carrier tube, is then introduced and the open end crimped to close it, leaving approximately 5 mm of carrier tube protruding to facilitate removal of the iridium wire.

All these techniques have been described in detail by many authors and the reader should refer to the following references for fuller details of the techniques used: Pierquin and Chassagne (1962), Pierquin (1964, 1966, 1971), Paine (1972), Pierquin *et al* (1978a), Hilaris (1975b) and Delclos (1982, 1984). Since these techniques are based on using wire or hair-pins of the same linear activity throughout the implant, new methods of dosimetry have been developed to calculate the dose within the treatment volume.

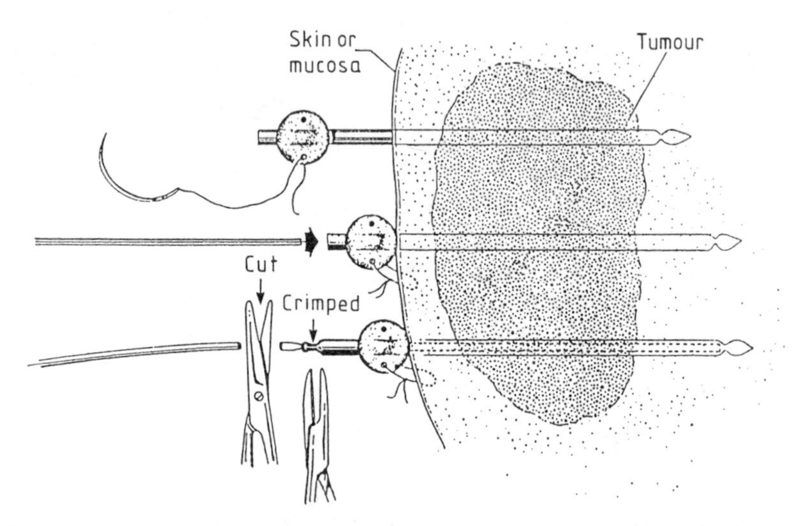

Figure 6.3 Insertion technique for tumours approached from one side. (Reproduced from Delclos (1984) by kind permission of Lea and Febiger, Philadelphia, USA.)

6.3.2 Source distribution rules of the Paris system

As with other interstitial treatment techniques, iridium wire implanted within tissue creates areas of very high dose in the immediate vicinity of the wire, which reduces rapidly with distance. In order to ensure adequate coverage of a lesion, various rules have been established based on the fundamental concept of a variable relationship between the length of the radioactive source and the spacing between each source length. These rules, detailed by Pierquin *et al* (1978a and b), have become known as 'The Paris System'. Within this system no attempt is usually made to cross the ends of an implant. The basic rules of the system are as follows (Pierquin and Dutreix 1966, Pierquin *et al* 1969, Marinello *et al* 1978, Dutreix *et al* 1979, 1982).

(i) The active sources should be arranged in parallel and straight lines where possible.
(ii) The spacing between lines should be as uniform as possible.
(iii) Each source should be a continuous length of linear activity radioactive wire and the linear activity of each line should be identical.
(iv) Each source should extend beyond the limits of the tumour volume so that the minimal dose calculated at the reference isodose is achieved over all the target volume.
(v) Although the spacing between sources of any one implant is equal, the separation between sources is dependent on the size of the volume, varying between 5 to 10 mm for a very small volume to 15 to 22 mm for a very large volume. These variations are necessary because of the different diameters in the isodose lines around wires of different lengths.
(vi) Due to the tolerance of different tissue to radiation, a smaller separation between active wires is required for tissues of lower tolerance. This has the effect of reducing the size of the high dose volume in the neighbourhood of the wire. On no account should the diameter of the high dose volume exceed 1 cm.
(vii) For parallel sources the dose calculation is made in a single plane perpendicular to the axis of implantation, passing through the centre of the volume.
(viii) The maximum activity in any implant using iridium wire should not exceed 150 mCi, i.e. approximately 90 mg Ra eq.
(ix) If more than one plane is used to cover the tumour volume, then the different radioactive lines should be implanted such that a wire in one plane is opposite a gap between wires in another plane.

Whilst these guide-lines will deliver a homogeneous dose to the target volume, dosimetric guide-lines are helpful to assist the placement of the sources within various size lesions. These guide-lines have been derived from clinical experience and calculated using computer techniques and are dependent on the shape of the implant.

6.3.2.1 Single wire implants. The dose variation around a single wire is dependent on its length. As the active length of the wire increases, it is found that the diameter of the circular isodose distribution around the wire also increases. This variation also explains why the spacing between lines in a multisource implant can be increased as the wire length increases. Based on the maximum acceptable area of high dose, the Paris system recommends limiting the width of the treated volume from a single wire to the distances shown in table 6.9. The length of wire is chosen so that adequate coverage is given to the target volume and the recommended 'useful' length treated is 0.7 times the radioactive length.

Table 6.9 Width of treated volume for various active lengths.

Length of active wire (mm)	10	20	30	40	50	60	70	80	90	100
Width of treated volume (mm)	9	12	13	14	14	15	15	15	16	16

6.3.2.2 Planar implants. With reference to figure 6.4, the dimensions of the volume are defined as follows.

(i) Length of treatment volume, l, is the smallest distance between the borders of the treated volume at either end of the volume, between the active wires and parallel to them. For single-plane implants this is measured in the plane of the implant but mid-way between the planes for more than one plane.

(ii) The thickness of the treatment volume, t, is the smallest distance between the two parallel planes which are tangents to the reference isodose at the boundary of the treatment volume, at either end of the volume at a point between the active wires.

(iii) The width, w, of the treatment volume is the distance between the boundaries of the treated volume in the axis perpendicular to the centre of the radioactive wires and in the plane of these radioactive sources. This width need only be considered in the calculation of single-plane implants.

(iv) The lateral margin, d, is the distance of the outer margin of the volume from two peripheral sources and is the minimum distance measured between the reference isodose and a line joining the points where any two sources intersect the central plane.

The variation in separation between adjacent sources is necessary in order to achieve greater dose uniformity. In the Paris system the general principle is that the longer the wire the greater the separation, with an upper limit of 2.2 cm. This corresponds to a maximum diameter of the high dose volume of approximately 1 cm around a 10 cm long source—the high dose volume being defined as that volume receiving twice the

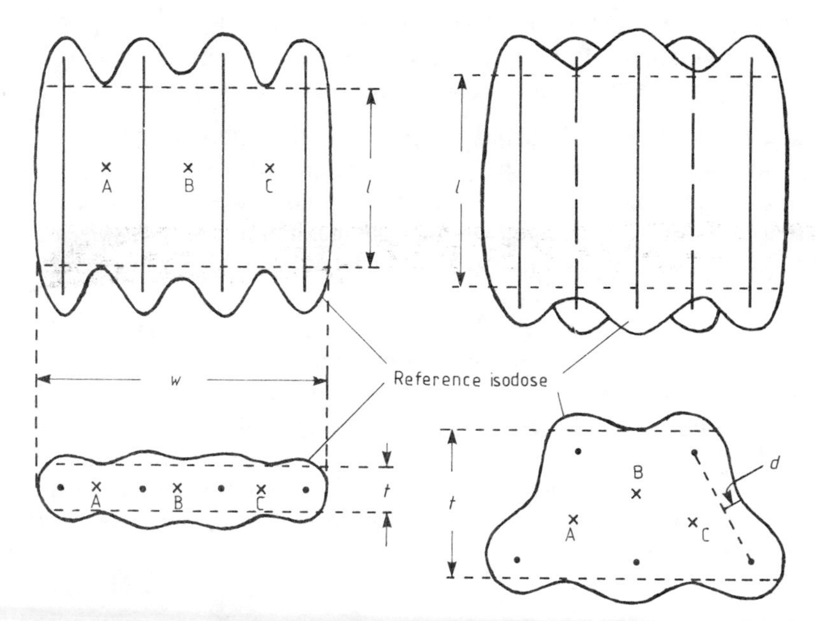

Figure 6.4 Positions of the basal dose reference points, reference isodose and dimensions of the volume defined for a single-plane and double-plane iridium-192 wire implant. (A cross indicates a basal dose calculation point.)

reference dose. This approach is justified on the basis that: (a) as the distance between the source and the reference isodose increases, the diameter of the high dose volume surrounding the sources increases proportionally less; and (b) the risk of creating hot or cold spots becomes greater with very small spacing.

Table 6.10 Relationships for planar implants.

Implant	l	w	t	d
Single plane	$(0.65–0.75)L$†	$(n-1)s + 2 \times 0.37s$	$(0.5–0.6)s$‡	$(0.34–0.37)s$
Triangular double plane	$0.7L$	–	$1.2s$	$0.15s$
Square double plane	$0.7L$	$(n-1)s + 2 \times 0.37s$	$1.5s$	$0.28s$

† Variation between 0.65 and 0.75 as the active length, L, increases from 10 to 100 mm.
‡ Variation between 0.46 and 0.59 as the active length L, increases from 10 to 100 mm.

If the separation between the sources is s and n is the number of wires of active length L, then the relationships relating s and L to l, w, t, and d are given in table 6.10 for planar implants.

6.3.2.3 Non-planar implants. Often it is clinically desirable to treat cylindrical volumes with straight and parallel radioactive sources. In these circumstances it is recommended that not more than five sources are used forming a pentagon and that the maximum spacing between the sources should not be more than 2.1 cm. With this configuration the treatment volume is almost cylindrical in shape, with the reference isodose having a diameter of 4 cm. This form of treatment is often achieved with radioactive hair-pins, single pins and loops of wire. When hair-pins are used (which means that the end is effectively crossed), the length, width and thickness of the basal dose volume are given by table 6.11 for various positions along the hair-pin. The useful length is defined as the length of the treated volume at a level where the thickness of the volume is 90% of that calculated at the centre.

Table 6.11 Parameters of basal dose volume.

Length of pin (mm)	Basal dose rate ($cGy\,h^{-1}$)	Volume width (mm)		Volume thickness (mm)		Volume length (mm)	
		Centre	Head	Centre	Head	Actual	Useful
20	36.2	20	18	6	9	15	11
30	38.2	20	17	7	9	24	18
40	39.9	21	17	8	9	33	25
50	41.1	21	17	8	10	42	35

6.3.3 Dose specification and calculation

In the Paris system the dosimetry is based on the basal dose rate (BD) which is a measure of the dose rate in the centre of the treated volume. It is calculated from the position of the sources in the central plane and is the minimum dose rate between a pair or group of sources. It must be recognised that the relationship between the basal dose and the spacing of the radioactive wires is critical. With a single hair-pin of standard linear activity of 1 $mCi\,cm^{-1}$ the basal dose varies according to length of pin and is about 40 $cGy\,h^{-1}$, as shown in table 6.11. For two parallel wires the variation in separation to achieve a constant basal dose rate with a different length of wire is shown in table 6.12.

For more than two wires the basal dose rate for the whole implant is taken to be the average of dose rates calculated at each minimum point. For various source configurations the basal dose rate is calculated as shown in figure 6.5. In the case of single-plane implants the points of minimum dose rate are at the mid-points between sources in the central plane, whereas for triangular implants the minimum points are where the perpendicular bisectors of the sides of the triangle intersect. It must be noted that in this dosage system rectangles and triangles with an obtuse angle are not acceptable for proper implant geometry.

Table 6.12 Spacing between two active iridium-192 wires of 1 mCi cm^{-1} linear activity to achieve defined basal dose rates in cGy h^{-1}. (Modified data from Pierquin *et al* (1978a).)

	Basal dose rate (cGy h^{-1})			
	40	35	30	25
Active length (mm)	Spacing between adjacent wires (mm)			
10	8	9	10	11
20	11	12	13	14
30	12	13	14	16
40	12	13	15	17
50	13	14	16	18
60	13	14	16	19
70	13	15	17	19
80	13	15	17	20
90	13	15	17	20
100	13	15	18	21

Having defined the basal dose, the treatment volume is defined as that volume enclosed by the reference isodose, which is 85% of the basal dose. This is the dose rate used for calculating the duration of the implant to give the prescribed dose to the tumour. The high dose volume is therefore that volume enclosed by the 170% isodose contour. It is evident that the shape of the reference isodose is irregular and care must be taken to ensure that this envelope encloses the target volume. Once the source positions within a tumour volume have been defined (by the use of orthogonal or stereo radiographs, transverse tomographs or visual inspection) the dose rate to each of the minimum dose points can be determined and hence the basal dose rate calculated. It is necessary to know the length of the wire and usually this is obtained from a knowledge of the wires inserted. In the special cases of a hair-pin or a

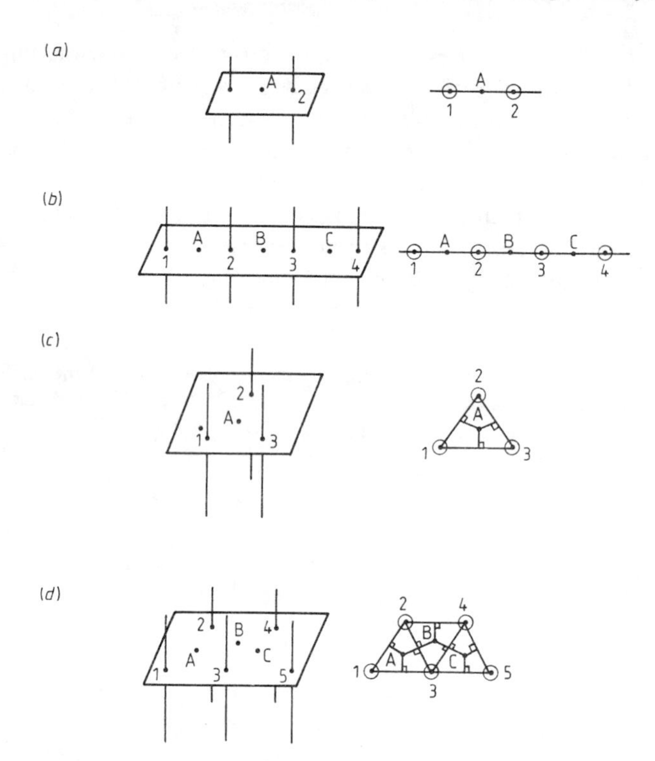

Figure 6.5 Calculation of points for the basal dose rate. (*a*) Single-plane implant, two parallel wires; basal dose rate = dose rate at A. (*b*) Single-plane implant, four parallel wires; basal dose rate = (dose rate at A + dose rate at B + dose rate at C)/3. (*c*) Three parallel wires forming a triangle. Basal dose rate = dose rate at A. (*d*) Five parallel wires in two planes. Basal dose rate = (dose rate at A + dose rate at B + dose rate at C)/3. (After Pierquin *et al* (1978a), reproduced by permission of Warren Green Inc., St Louis, Missouri.)

wire inserted as a loop, the value taken for the active length is half of the total length of the wire which makes up the hair-pin or loop. Therefore, a hair-pin of length 5 cm has a total length of 11.2 cm and can be considered as two separate wires of active length 5.6 cm provided the length of the loop or hair-pin is more than twice the separation of the individual wires (Pierquin *et al* 1978b) and the wires are parallel to each other. A similar approximation is applied to determining the active length of a single pin. The calculation of the basal dose is then made using one of the following alternative methods.

(1) Knowing the distance from each line source to the points of calculation, the exposure rate due to each source can be determined using the cross-plot charts derived by Hall *et al* (1966b): it can be subsequently converted to read absorbed dose in tissue (Welsh *et al* 1983), as discussed in §3.6.2.3. These charts are based on a linear activity of 1 mg Ra eq cm^{-1}. This method is often referred to as the Oxford iridium-192 system.

(2) The method, adopted by Pierquin in Paris, makes use of 'escargot curves' (figure 6.6). To use these curves it is necessary to reconstruct the source distribution in the central plane, typically by using a tomogram. The escargot curves, described by Schlienger *et al* (1970), are constructed with the appropriate magnification factor so that they can be placed directly onto the reconstruction, the centre of the curve being placed on each point where the source intersects the central plane. The graph is then turned so that the line appropriate for the length of the active wire passes through the point at which the dose is required. The dose rate in Gy h^{-1} is then read directly from the spiral line. These curves were initially based on linear activity of 1 mCi cm^{-1}, although a new method of specifying source activity was adopted in 1974 following the proposal of Dutreix (1974).

In both the above methods, once the dose has been determined from each source to each calculation point, the basal dose is found by taking the mean of the values obtained at all the points of calculation. A correction has then to be applied to determine the actual basal dose rate to account for the actual linear activity of the wire used and also for the fact that the activity of the wire will decay during the course of the implant. The time for which the therapy is to be given is then determined from the reference dose rate.

Recently, Casebow (1984) has produced tables for standard iridium-192 wire single and double implants which enable the wire activity required for a given implant to be readily determined, together with the dimensions of the treatment volume within the 85% dose contour. This work made no correction for oblique filtration and assumed that the absorption and scattering within tissue could be ignored. Further analysis extended this approach to various source arrays and also included a correction for oblique filtration (Casebow 1985).

In interstitial techniques, however, an 'ideal' implant is rarely achieved and often the implanted wires do not comply with conditions of accept-

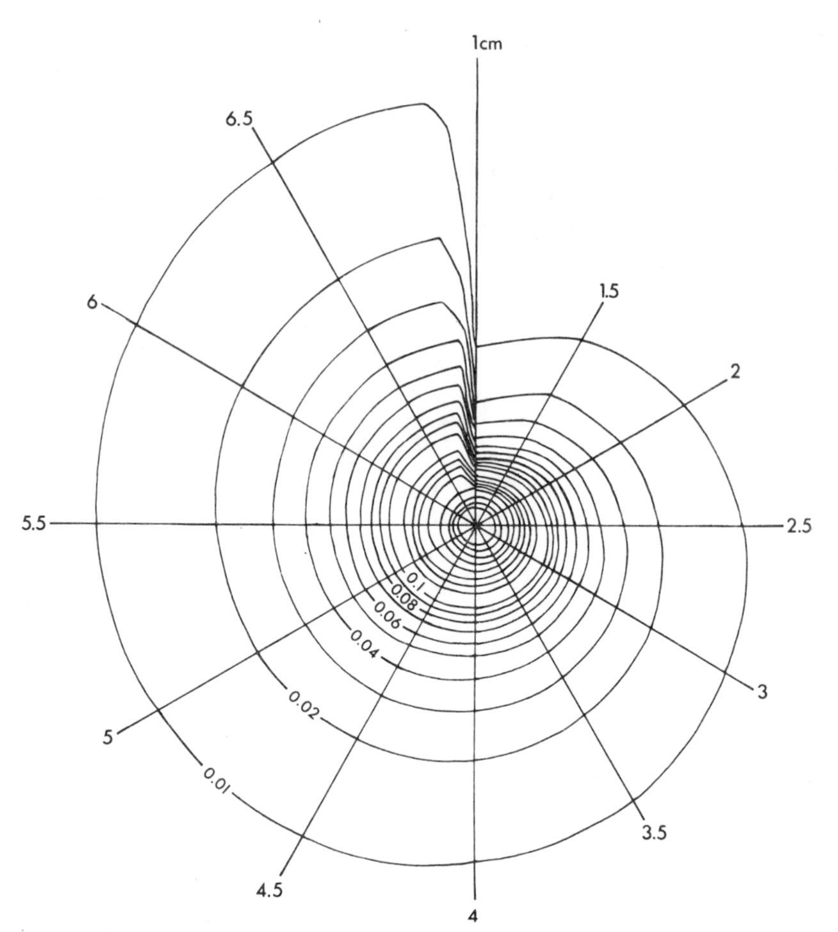

Figure 6.6 The 'escargot' curve. (The figures on the curved lines give absorbed dose rate in Gy h^{-1}.) (Kindly supplied by Amersham International plc with permission from *Acta Radiologica*.)

able geometry for the Paris system, i.e. no obtuse angles. This leads to a consequent increase in the size of the 'hot spot' areas if the dose is solely determined in terms of the basal dose rate. To overcome the problems associated with non-ideal implants, Mayles *et al* (1985) have developed a dose optimisation method which achieves a clinically acceptable dose distribution by using sources of different linear activity rather than sources of the same linear activity, as prescribed by the Paris system, or by removing sources of the same linear activity at different times. For this system to work it is necessary to calculate the treatment dose rate and this is done by making slight modifications to the Paris method of determining the reference points. In this proposed system the authors have suggested the following.

(i) Where a section of the implant is better described by a square or quadrilateral whose diagonals do not differ in length by more than 10%, the reference point should be the centre of the quadrilateral.
(ii) Where a triangle contains an obtuse angle, the mid-point of the side of the triangle opposite the obtuse angle should be chosen as the reference point.

Whilst these do not conform to the recommendations of the Paris system, the authors have shown that provided the dose distribution is calculated and is satisfactory, then the change is not clinically significant. Once the dose rate at the reference points is calculated, the 'treatment dose rate' can be taken as 85% of the average value of the dose rate at the reference points, the basal dose rate.

To achieve a method of determining the optimum dose distribution for unevenly spaced wires and for wires of different linear activity, Mayles *et al* (1985) sought an alternative criterion to the 85% of basal dose rate (as used in the Paris system) for use as the treatment dose rate. In this study the aim was to produce as uniform a dose distribution as possible while limiting the treatment dose according to the size of the high-dose region created. From an examination of the dose distribution for ideal geometry implants, a measure of dose uniformity in terms of the proportion of the treatment volume or area in the basal plane which reached dose levels of given percentages above the treatment dose was calculated. This study showed that the calculation of areas of very high percentage of treatment dose was not accurate, whilst the value of the line enclosing 50% of the treated area in the basal plane was approximately constant for all the various ideal implant geometries and wire spacing. This led them to define the treatment dose rate with the criterion that 'not more than 50% of the treated area in the basal plane of the implant shall be given a dose greater than 127% of the prescribed dose'. Using this criterion it was possible to produce clinically acceptable dose distributions using wires of different linear activities for implants which under the Paris system would have been deemed to be unacceptable (figure 6.7).

In an attempt to assess the differences between the Paris and Manchester dosage systems, Gillin *et al* (1984) carried out a comparison for the two systems for single- and double-plane implants. For the examples considered it was found that the Paris system used considerably fewer sources, which resulted in larger volumes of high dose than produced by the Manchester system. Even though the two systems differ in the way they define the prescribed 'dose', the dose rate achieved when using the Paris system source strengths of 4.2–6.4 μGy h^{-1} cm^{-1} m^2 in air (1.0–1.5 mCi cm^{-1}) is comparable to the 10 Gy per day of the Manchester system.

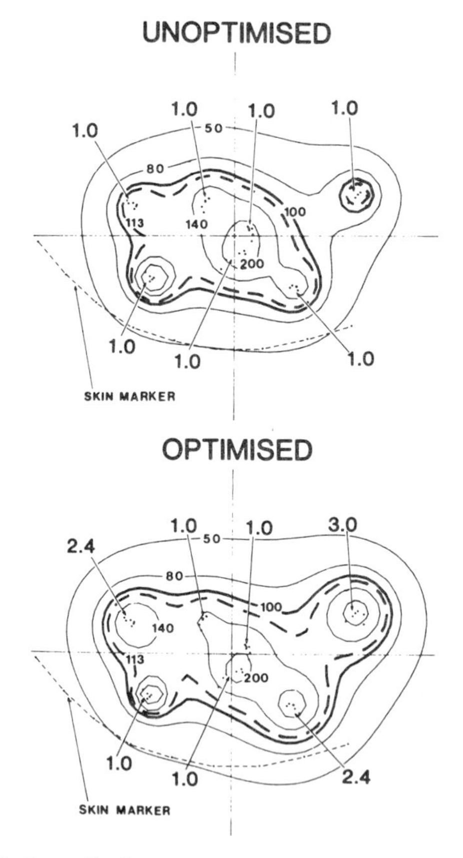

Figure 6.7 Dose distributions from an implant. Numbers next to wires indicate the relative activity of wire. 100% line defines treatment isodose line and dashed line, 113%, is the reference isodose line of the Paris system. (After Mayles *et al* (1985), reproduced by permission of *British Journal of Radiology*.)

6.3.4 Clinical examples

In calculating the following examples, the Oxford cross plots (which give the dose rate in $cGy\,h^{-1}$ with distance from the wire for various cross plots for an equivalent activity of 1 $mg\,Ra\,eq\,cm^{-1}$ of wire) have been used for wire where the source strength has been defined in terms of exposure rate or air kerma rate at 1 m. The assumption has been made that the implant is ideal and that the plane of calculation is through the centre of all the wires or hair-pins.

6.3.4.1 Single-plane implant. A breast lesion is being treated with five iridium wires, each 10 cm long. If the spacing between the wires is 1.5 cm and the linear activity of the wire is 0.55 $mR\,h^{-1}\,cm^{-1}$ at 1 m three days prior to the insertion, calculate the basal dose rate and the time required to give 60 Gy to the reference isodose.

Using the cross plots the basal dose rates at the four reference points, mid-way between each pair of wires are as shown in table 6.13.

Table 6.13 Calculation of basal dose rates for single-plane implant.

	Basal dose point							
	A		B		C		D	
Wire	Distance (cm)	Dose rate ($cGy\,h^{-1}$)	Distance (cm)	Dose rate ($cGy\,h^{-1}$)	Distance (cm)	Dose rate ($cGy\,h^{-1}$)	Distance (cm)	Dose rate ($cGy\,h^{-1}$)
1	0.75	29.0	2.25	8.2	3.75	3.9	5.25	2.3
2	0.75	29.0	0.75	29.0	2.25	8.2	3.75	3.9
3	2.25	8.2	0.75	29.0	0.75	29.0	2.25	8.2
4	3.75	3.9	2.25	8.2	0.75	29.0	0.75	29.0
5	5.25	2.3	3.75	3.9	2.25	8.2	0.75	29.0
Total dose rate ($cGy\,h^{-1}$)		72.4		78.3		78.3		72.4

Average BD for 1 $mg\,cm^{-1}$ = (72.4 + 78.3 + 78.3 + 72.4)/4 = 75.35 $cGy\,h^{-1}$. Linear activity at the beginning of implant = (0.55/0.825) $\times$ 0.972 $mg\,Ra\,eq\,cm^{-1}$ = 0.648 $mg\,Ra\,eq\,cm^{-1}$. Therefore BD at the beginning of implant = 75.35 $\times$ 0.648 = 48.83 $cGy\,h^{-1}$. The dose rate at the reference isodose = 48.83 $\times$ 0.85 $cGy\,h^{-1}$ = 41.50 $cGy\,h^{-1}$. Thus, *effective treatment time* for 6000 cGy = 6000/41.50 = 144.6 hours.

To achieve this effective time, the duration of the implant can be determined by interpolation from table 5.7 or from equation (5.2) such that

$$T = \log(1 - \lambda T_{\text{eff}})\lambda^{-1}.$$

Thus the actual duration of this implant needs to be 148.8 hours = 6 days 5 hours.

To a first approximation the time can be determined by calculating the BDR at the approximate mid-point of the treatment. This obviates the need to determine the actual treatment time from the effective time.

6.3.4.2 Double-plane implant. A tongue lesion is treated with three hair-pins, each 4 cm long and of 0.335 μGy h^{-1} mm^{-1} linear activity at 1 m on the calibration data. The tomographic cross section through the central plane is as shown in figure 6.8, together with the basal dose reference points. For calculation purposes the hair-pin is considered as two separate wires of length 4.6 cm, as discussed in §6.3.3. See table 6.14 for calculations.

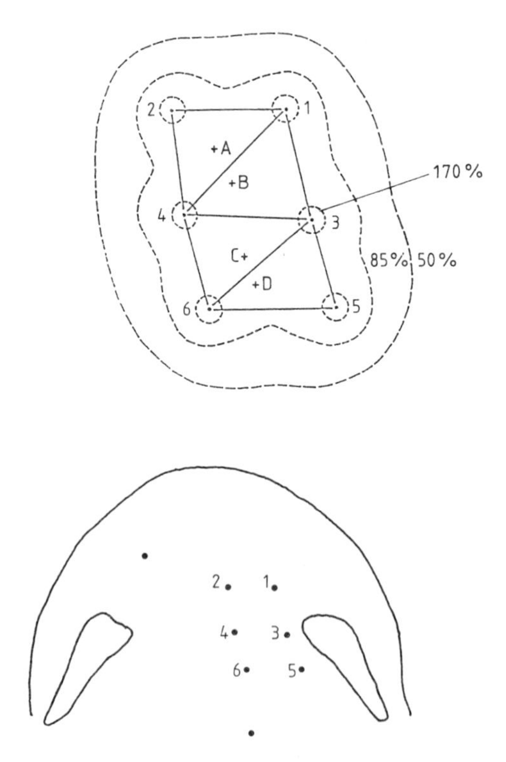

Figure 6.8 Tomographic section through a three hair-pin tongue implant with resulting dose distribution.

Table 6.14 Calculations of basal dose rates for double-plane implant.

	Basal dose point							
	A		B		C		D	
Wire	Distance (cm)	Dose rate ($cGy\,h^{-1}$)	Distance (cm)	Dose rate ($cGy\,h^{-1}$)	Distance (cm)	Dose rate ($cGy\,h^{-1}$)	Distance (cm)	Dose rate ($cGy\,h^{-1}$)
1	0.80	24.4	0.80	24.4	1.60	9.6	1.70	8.9
2	0.80	24.4	1.05	17.1	1.90	7.4	2.00	6.8
3	0.95	19.6	0.70	28.5	0.80	24.4	0.80	24.4
4	0.80	24.4	0.75	26.3	0.85	22.5	1.00	18.4
5	1.85	7.7	1.60	9.6	0.90	21.0	0.70	28.5
6	1.95	7.0	1.75	8.4	0.85	22.5	0.85	22.5
Total dose rate ($cGy\,h^{-1}$)		107.5		114.3		107.4		109.5

Average BD for $1\ mg\,cm^{-1} = (107.5 + 114.3 + 107.4 + 109.5)/4 = 109.7\ cGy\,h^{-1}$. The date of the middle of the insertion is seven days after the calibration date, thus the linear activity $= (0.335/0.720) \times 0.937 = 0.436$ mg Ra eq. The average BDR $= 0.436 \times 109.7\ cGy\,h^{-1} = 47.8\ cGy\,h^{-1}$. Dose rate at reference isodose $= 47.8 \times 0.85 = 40.6\ cGy\,h^{-1}$. Duration of implant for 60 Gy = 147.6 hours, i.e. 6 days 4 hours.

6.3.5 Dosimetry of seed implants

Whilst the dosimetry of the Paris system is based on the use of constant linear activity wire, the after-loading technique developed by Henschke *et al* (1963) uses nylon ribbons loaded with iridium-192 seeds. These seeds typically have a core of Pt–Ir alloy which is either doubly encapsulated in stainless steel (Suthanthuran 1981) or sheathed in 0.1 mm thick platinum (Stephens 1981) which tends to give them a 20% anisotropy in photon fluence (Ling *et al* 1983a). In clinical use the seeds are supplied in nylon tubes spaced at 1 cm intervals, usually with twelve seeds per tube. Implanting seed trains means that an array of seeds of equal activity at defined spacing is produced which satisfies neither the Paris nor Manchester dosage systems. A dosage system has therefore been devised by Anderson *et al* (1981b) which is based on the Memorial dosage system devised by Laughlin *et al* (1963). In this system (Anderson 1983) a rectangular array of uniformly spaced ribbons is assumed and guide-lines are given for pre-implantation planning which enables the radiotherapist to determine the required source spacing and activities based on the use of a nomogram. From a knowledge of the length and width of the region to be implanted it is possible, for planar implants, to determine the number of ribbons required to achieve a dose of 10 Gy per day at a 'peripheral' reference point. This point is located in a plane 5 mm from the source plane and at a point 1.5 spacing units (15 mm) inwards from the end source along the ribbon direction, and 0.5 spacing units (5 mm) inward from the transverse edge ribbon.

Experience has shown that the 'peripheral' dose determined from the Memorial planar guide (Anderson 1983) and the dose determined from the Manchester planar implant table are in good agreement. Whilst the Memorial planar guide-lines only strictly apply to single-plane implants, Kwan *et al* (1983) have demonstrated that although the optimal source and ribbon spacing for a single-plane implant is 10 mm, for double-plane implants the preferred ribbon and planar separation is 15 mm, maintaining the sources at 10 mm intervals in the ribbons.

In a comparison of the dosimetry associated with iridium wire and seed ribbons, Marinello *et al* (1985) have demonstrated that whilst seed ribbons produce greater inhomogeneity and a loss of flexibility, they can nonetheless be used in the Paris system provided certain conditions are satisfied. From the study, the authors showed that provided seed ribbons are chosen in which the ratio between the seed centres and the length of seed is less than or equal to 1.5, the Paris system remains valid and the same source-type geometric relationships exist as for continuous lines of iridium wire.

6.4 Permanent Seed Implants

6.4.1 Dosage systems

The two main radium dosage systems, whilst being initially developed for use with radium needles, have also been extensively used with radon-222 seeds and more recently with gold-198 seeds. With the Paterson–Parker system the seeds have to be placed in accordance with the distribution rules, but this is not always feasible, particularly in small treatment volumes. It is therefore often more convenient to space equal activity seeds at a nominal spacing of say 1 cm, thereby achieving a uniform distribution of sources in accordance with the Quimby system, rather than the Paterson–Parker system. Various authors (Shalek *et al* 1957, Laughlin *et al* 1963, Dutreix 1961, Henschke and Cevc 1968) have therefore considered other dosage systems for use with seed implants based on distributing the seeds ideally as a lattice.

6.4.1.1 Shalek dosage system. Shalek *et al* (1957) derived a dosage system by considering various sizes of cubic and spherical lattice seed spacing, the spheres being formed from cubes by the removal of the corner and edge seeds. The stated dose in this system is similar to that of the Paterson–Parker system, ie. 10% above the absolute minimum within the volume. The approach adopted was to calculate the dose variation along a 'low-line', which is the line which includes the minimum dose rate for a given distance from the centre, and then to verify experimentally the calculated value which made *no* allowance for photon absorption or scattering in tissue.

For cubes the low-line was taken to originate at the centre of the cube and pass through the central face, the doses along this line being calculated by considering each seed as a point source. This procedure was adopted for cubic and spherical seed lattices and compared with the low-line derived for a cylindrical needle implant complying with the Manchester distribution rules. For the spheres the low-line passes through the point of minimum dose on the surface, whereas for cubes the point of minimum dose is not on a low-line since it is near the corner of the cube. Using this approach, Shalek showed that the ratio of the lowest surface dose to the centre varied from 0.61 to 1.00 for cubes and is approximately 0.62 for spheres. This compares with a ratio of 0.80 for Paterson–Parker cylinders; thus the dose variation within a volume implanted on a 1 cm cubic lattice is somewhat greater than across a Paterson–Parker needle implant. This work resulted in a dosage table for volume implants using radioactive seeds. Table 6.15, giving the mg h

Table 6.15 Number of mg h for the Shalek *et al* (1957) system for seed implants. (Modified data.)

Volume (cm^3)	Number of mg h per 10 Gy	Volume (cm^3)	Number of mg h per 10 Gy
1	89.9	30	421
2	113	40	491
3	131	50	558
4	147	60	622
5	158	70	684
10	232	80	744
15	292	90	803
20	343	100	858
25	382		

per 10 Gy, has been derived from the original work of Shalek *et al* using their recommended factor of 0.90 to convert roentgens to rads (Stovall and Shalek 1968). A comparison of these results with those of the Paterson–Parker and Quimby systems shows that for small volumes the percentage difference is considerable, i.e. 50% for 5 cm^3 volume; but for volumes greater than 100 cm^3, the differences are less than 5%.

6.4.1.2 Dutreix dosage system. A further seed implant dosage system was also evolved by Dutreix (1961) for use with gold seeds. In this approach variation in lattice spacing was allowed. The main assumptions made in this analysis were as follows.

(i) A sphere of radius r containing a lattice of point sources may have its distribution of exposure represented by that of a sphere of the same radius having a continuous radiation density, both spheres having the same radioactive content.
(ii) The sphere containing a lattice of point sources may be modified so as to be regarded as a cuboid containing the same number of point sources, N, as the sphere. For practical purposes the number of seeds used should be less than 30.

By considering the dose rate at the centre of a sphere of radius, r, and equating it to the activity of a cuboid lattice containing N point sources, Dutreix derived a relationship between the number of seeds, N, each of activity A, and the volume of radius r to yield an exposure of 10 000 R of the form

$$N = 14.93\, r^2/A.$$

These assumptions, however, led to large differences when compared to other workers.

6.4.1.3 Laughlin dosage system. In both of the above systems no account was taken of the build-up factor. However, Laughlin *et al* (1963) considered this effect when making a number of calculations for a large variety of seed implants in the shape of planes, cubes and oblongs. Their dosage system determined dose distributions in intermediate planes and isodose planes. They used a stated dose described in terms of the following.

(i) The reference maximum dose is the highest dose found in any set of isodose planes.
(ii) The minimum peripheral dose is the lowest dose found at the

intersection of the periphery of each seed array with the isodose planes.

(iii) The centre line peripheral dose is the volume of the dose at the intersection of the centre line (a line included in that isodose plane containing the reference maximum through it, parallel to an edge of the implant) and the periphery of the implant.

The tabulated results are in terms of the number of mg h required to produce 10 Gy in water at the location of the minimum peripheral dose (table 6.16); the centre line peripheral dose and reference maximum dose are given as a ratio of the minimum peripheral dose for both area and volume implants. A factor of 0.965 was used to convert exposure to

Table 6.16 Number of milligram–hours required to deliver 1000 cGy at designated points for area and volume implants. (Modified data from Laughlin *et al* (1963).)

Area implants			Volume implants			
Area (cm^2)	mg h for 10 Gy at minimum peripheral dose, Point A	Reference maximum dose as ratio of *A*	Volume (cm^3)	mg h for 10 Gy at minimum peripheral dose, Point B	Centre line peripheral dose as ratio of *B*	Reference maximum dose as ratio of *B*
1	95	1.13	1	95	1.00	1.13
2	116	1.23	5	184	1.12	1.27
3	138	1.29	10	254	1.19	1.37
4	165	1.33	15	322	1.22	1.43
5	189	1.37	20	378	1.26	1.48
6	211	1.39	25	433	1.28	1.52
7	236	1.42	30	472	1.29	1.55
8	256	1.44	40	560	1.32	1.60
9	280	1.45	50	640	1.34	1.64
10	300	1.47	60	719	1.35	1.67
12	342	1.50	80	865	1.38	1.73
14	380	1.52	100	1000	1.40	1.77
16	419	1.54	120	1130	1.41	1.81
18	454	1.56	140	1240	1.42	1.83
20	488	1.57	160	1365	1.43	1.86
25	570	1.61	180	1475	1.44	1.88
30	650	1.63	200	1585	1.45	1.90
35	734	1.66	250	1840	1.47	1.95
40	812	1.67	300	2080	1.48	1.98
45	892	1.69	350	2310	1.49	2.00
50	973	1.71	400	2540	1.50	2.20

absorbed dose. To calculate the number of seeds, N, to be inserted into a given volume use was made of the following expression

$$N = (L + 1)^3$$

where L is the length of the side of the cube.

It will be noted that these data differ from those of Shalek and Paterson–Parker, due in part to the differences in the point of dose specification, in specific gamma-ray constant and the inclusion of the build-up factor and the roentgen-to-rad factor, as well as differences in source distribution with respect to the Paterson–Parker system. Mould (1966) has compared these various tables and shown that by applying a correction of 1.15 in an attempt to account for the differences in dose specification mentioned, the data of Laughlin and Shalek are within 2% at 1 cm^3 and 8% at 125 cm^3.

6.4.2 Dimension averaging system

This system is an extension due to Laughlin and is directed towards obtaining a certain minimum tumour dose regardless of tumour volume. In this method a defined activity is implanted which is directly proportional to the average of the three mutually perpendicular dimensions of the implant region, a, b, and c. In mathematical terms, the activity is given by

$$A = Kd_a \tag{6.3}$$

where the 'average dimension', d_a, is given by $d_a = (a + b + c)/3$ and K is the constant of proportionality.

This method was first used with radon seeds and clinical experience led to a value for the constant of proportionality, K, of 10 for radon-222. The dose–volume relationship implied by the dimension averaging was developed by Henschke and Cevc (1968) and made use of the Quimby volume implant data. The dose, D, in cGy may be written

$$D = CMV^{-1/2} \tag{6.4}$$

where the constant $C = 10$, V is the volume and M is the cumulative activity in mCi h.

For a spherical implant the average dimension is the diameter: thus

$$V = (4/3)\,\pi\,(d_a/2)^3 = (\pi/6)\;d_a^3.$$

Therefore

$$d_a = 1.24\;V^{1/3}. \tag{6.5}$$

Since $M = AT_{ave}$, where T_{ave} is the average life of the isotope, then combining equations (6.3), (6.4), and (6.5) we have a relationship which defines the dose in terms of activity and volume such that

$$D = 1.24CKT_{ave}V^{1/6} \quad \text{cGy}. \tag{6.6}$$

However, in addition to the total activity required, it is necessary to know (a) how many seeds to implant and (b) how far apart to space them in order to achieve the desired distribution of seeds throughout the defined volume. Anderson (1976) derived relationships for these parameters such that the volume for a sphere is given by

$$V = (\pi/6)f_e^3 d_a^3 \tag{6.7}$$

and for a cylinder by

$$V = (\pi/4)f_e^3 d_a^3 \tag{6.8}$$

where f_e is an elongation correction which for a sphere and a right circular cylinder is equal to one.

The volume v_s allocated to each seed of activity q is given by

$$v_s = Vq/A. \tag{6.9}$$

By substituting for A in equation (6.3) and V in equation (6.7) the volume for a spheroidal region is

$$v_s = (\pi/6K)f_e^3 d_a^2 q. \tag{6.10}$$

The spacing between seeds, u_s, is taken to be the cube root of the volume per seed, on the assumption that the seeds form a cubic lattice within the region, thus

$$u_s = (\pi/6K)^{1/3} f_e^{2/3} d_a^{1/3} q \tag{6.11}$$

and for a cylindrical volume, the term in the bracket becomes $(\pi/4K)$. The elongation correction, f_e, is derived in terms of the elongation factor r, the ratio of the longest to shortest dimension. For oblate spheroids and right circular cylinders whose length is less than the diameter

$$f_e = 3r^{2/3}/(2r + 1)$$

whereas for prolate spheroids and cylinders of length larger than the diameter

$$f_e = 3r^{1/3}/(r + 2).$$

It must be emphasised that any comparison between dosage systems must be done with care, since the points of calculation of the stated dose are not the same for each system.

6.4.3 Gold grain implants

Radioactive gold grains were first used for permanent implant by Colmery in 1951 and in the following year a technique for implantation using a gold grain gun was reported by Hodt *et al* (1952). The Mark 1 gun was loaded by unscrewing the needle and inserting a magazine of sources into the needle chamber, but this was susbsequently modified (Jones *et al* 1965) to make loading simpler. In this technique curved or straight needles are placed into the lesion and, using the graduation along the outside of the needle, a seed is inserted and then the gun (figure 6.9) withdrawn a given distance and another seed deposited and so on, the seed distribution being arranged to comply with one of the distribution rules mentioned above.

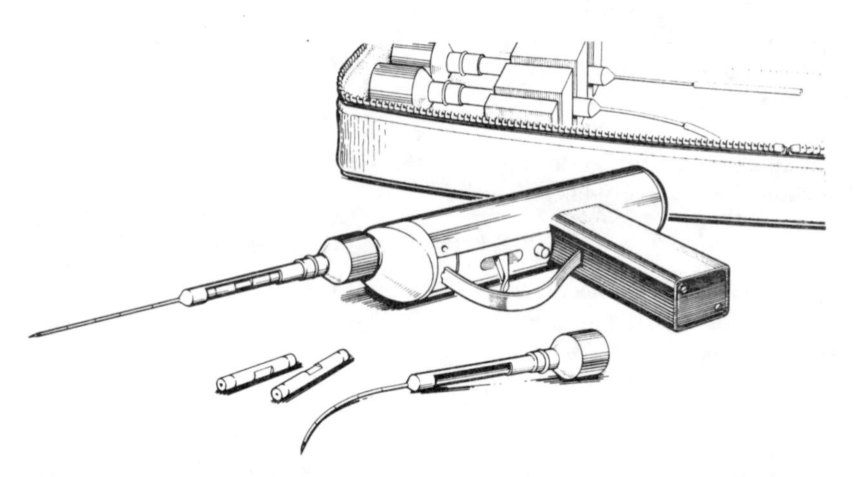

Figure 6.9 Implantation gun for gold seeds. (Reproduced by kind permission of Amersham International plc.)

In determining the number of seeds required in such a treatment, as shown in the following example for gold-198, it is necessary to determine the effective treatment time. From table 5.6 we know that this time is $t_{eff} = 93.4$ hours, that (from table 5.7) the cumulative dose in mg Ra eq h per initial activity is 0.719 (26.59) (MBq (mCi)). In addition, the biological factor must be remembered when prescribing dose.

Example 6.2
A permanent implant is to be performed to treat a lesion of 2.5 cm diameter in the floor of the mouth to give a dose of 55 Gy. What activity of gold-198 will be required?

Area of lesion $= \pi(2.5/4)^2 = 4.91\ \text{cm}^2$. From table 5.1, by interpolation, the number of mg h required for 10 Gy = 171.8 mg h. The activity of gold-198 required to give a dose of 55 Gy is $(171.8/0.719) \times 5.5 =$ 1314 MBq or $171.8/26.59 \times 5.5 = 35.54$ mCi.

Using the volume implant rules, 2/3 of the activity is placed at the periphery and the remaining activity spread evenly over the central region. Since the circumference is 7.85 cm and the spacing between seeds can be 1 cm, eight seeds are to be placed on the periphery and four seeds in the centre, the apparent activity of each seed being 110 MBq (3 mCi).

For seed strength specified in air kerma rate at 1 m, then we require $171.8 \times 5.5 \times 7.2\ \mu\text{Gy h}^{-1}$ h at 1 m. Since the effective treatment time is 93.4 hours, then the total source strength required is $171.8 \times 5.5 \times 7.2/93.4\ \mu\text{Gy h}^{-1}$ at 1 m $= 72.8\ \mu\text{Gy h}^{-1}$ at 1 m. Thus the source strength of each of the twelve seeds should be $6.07\ \mu\text{Gy h}^{-1}$ at 1 m.

Calculations taking into account photon attenuation and scattering in tissue have shown that the uniformity of dose within a planar implant can be improved by modifying the Paterson–Parker rules (Dale 1976). By considering a dose matrix, Dale derived the optimum distribution for circular implants ranging in diameter from 2 to 6 cm by calculating the standard deviations of the doses in the matrix, expressed as a percentage of the average dose to all sites and then he used this as a measure of the uniformity for a given distribution. From a study of various loading distributions to the outer and inner rings and the centre spot, he derived various tables for different seed activities and numbers for different diameters of circles. The results of this study are shown in table 6.17 for circles in the range 2–6 cm using seeds of approximately 3 mCi activity or $6.2\ \mu\text{Gy h}^{-1}$ at 1 m. It should be noted that although the seed arrangements are different from those specified by the Manchester rules, the required activity per seed to deliver 60 Gy is in good agreement with that predicted by the Manchester radium dosage system. From this table it is seen that if 3 mCi seeds were used in the above example, eleven seeds would be placed on the periphery and one seed as a central spot; this is almost equivalent to using the planar implant rules for the distribution.

When using 4 mCi grains, Dale found that due to a reduced number of grains the percentage standard deviation was typically of the order of 7.5.

Table 6.17 Loading systems to give the *minimum* dose variation in a plane 5 mm from a plane containing gold grains, each of approximately 3 mCi (111 MBq) activity. (Modified data from Dale (1976.)

Characteristic	Diameter of circle (cm) 2.0	2.5	3.0	3.5	4.0	4.5	5.0	6.0
Number in outer ring	8	11	11	13	15	18	20	26
Number in 1/2 diameter ring	0	0	4	5	6	5	7	10
Number in centre	1	1	0	0	0	1	1	1
Number of mCi per grain per 60 Gy	3.02	3.03	2.79	2.83	2.90	3.04	2.96	2.90
Percentage standard deviation	3.12	3.58	5.01	4.67	4.89	5.27	5.26	7.13

6.4.4 Yttrium-90 implants

Certain disorders of the pituitary gland can be treated by implantation techniques. Yttrium-90, a beta-radiation emitter whose maximum energy is 2.2 MeV, can deliver a dose of 1000 Gy or more to the whole gland, whilst at the same time giving a dose of only 100 Gy beyond the gland periphery. At this dose level complete pituitary ablation occurs. In order to plan implants it is necessary to have depth dose data from yttrium-90 rods. Jones *et al* (1963) have described a method of using x-ray film as a radiation detector for measuring this depth dose within Mix D wax which acts as as a tissue-equivalent absorber. The film is calibrated with a standardised strontium-90 plaque. The rods implanted are commonly 6 or 7 mm long and have an activity of between 2 and 4 mCi. The activity of the rods is determined from equation (2.1) in §2.2, but it is desirable to check this prior to implantation. This has been done using the beta-window of a re-entrant ionisation chamber, NPL type 1383.

The depth dose measurements for various lengths of rod have been calculated by Jones *et al* (1963) together with isodose curves. These have been shown to be elliptical in shape and can be approximated by the equation

$$[x/(L/2 + D_{\mathrm{EO}})]^2 + [y/(0.5 + D_{\mathrm{BS}})]^2 = 1$$

where x and y are the distances from the centre of the rod, L is the rod length, D_{EO} is the distance from the surface of the rod measured along the major axis and D_{BS} is the distance from the surface of the rod measured along the minor axis.

6.4.5 Iodine-125 implants

Iodine-125 seeds have been in clinical use for permanent interstitial therapy since 1965. Much of the early work, both clinical and dosimetric, was done at the Memorial Sloan Kettering Cancer Center, New York, by Hilaris and colleagues. This Center has developed an after-loading technique based on that developed earlier for permanent implants using radon-222, gold-198 and iridium-192 (Henschke *et al* 1963). Iodine-125 has now almost totally replaced the use of these other seeds at that Center. Its long half-life (60 days) and low photon energy (28–35 keV) give distinct advantages, both in terms of radiation protection and easier source availability, since their effective shelf life is of the order of two months which makes them available for use at relatively short notice. In comparison with other implant seeds, the half value layer is much lower, 0.025 mm in lead compared with 4.5 mm for gold-198, and 2.0 cm in tissue compared with 6 cm in tissue.

The implant techniques developed by Henschke (1956) involve the introduction of standard Number 17 hollow needles into the tumour volume, into which the iodine-125 seeds are inserted. The needles are spaced about 1 cm apart for tumours of up to 4 cm in diameter and at larger separations for larger volumes. Various methods have been developed for introducing the seeds which are spaced into the needle so that the overall effect is to create a seed lattice (Scott 1972a and b, 1977).

The sites which have been treated by this technique have included the prostate (Hilaris *et al* 1974, 1975a, b, 1976, 1977), lymph nodes in the neck and unresectable lung tumours, all of which are discussed by Hilaris (1975b). Studies have suggested that iodine-125 seeds have a higher therapeutic ratio than either radon or gold seeds (Kim and Hilaris 1975), although the radiological basis for such superiority is not fully understood.

Two of the three possible explanations which have been advanced to account for this are the low dose rate, sustained over a long period of time, and a possible high RBE for the low-energy x-rays from iodine. Another factor may be that during the time of the implant changes can occur in the treatment volume and this is particularly significant if rapid tumour regression occurs during an early stage after implantation, since this will result in a higher dose being given to the tumour volume than initially estimated (Tokita *et al* 1980).

The initial dosimetry associated with this technique was based on the dimension averaging technique discussed in §6.4.2. Due to the relatively long half-life, the dose achieved is much higher than normally given by permanent implant, of the order 160–200 Gy. Early clinical experience was obtained using a value of $K = 5$ in equation (6.3), irrespective of volume size. This meant that there was a gradual decrease in dose with increasing volume. However, it was observed that the peripheral implant dose fell off more rapidly than $V^{-1/6}$, in part due to the low energy of the x-rays. This led to a modification in the value adopted for K in order to increase the activity specified for larger volumes (Kuan and Anderson 1977). Thus the activity required is now determined from the following relationship:

$$A = 5d_a \qquad \text{for } d_a < 2.4 \text{ cm}$$

$$A = 3.87 d_a^{1.293} \qquad \text{for } 2.4 \text{ cm} < d_a < 3.24 \text{ cm}$$

and

$$A = 2.76 d_a^{1.581} \qquad \text{for } d_a > 3.24 \text{ cm}.$$

A subsequent modification was made to the Memorial protocol in 1979 when an additional 1 cm was added to the measured average dimension. The effect of these changes to the minimum peripheral dose and maximum central dose has been studied by Rao *et al* (1981). To aid rapid calculation Anderson (1975) devised a nomogram which related the total activity required to the average dimension and the seed spacing. This original nomogram was subsequently modified following the recommendations of Kuan and Anderson (1977). In practice, this nomogram is usually available at the time of implantation in order that the desired seed and needle spacing can be determined to match the available sources and size of tumour.

On completion of the implant procedure, radiographs are taken to determine the spatial distribution of the sources, the technique usually adopted being that of taking stereo-shift films (Chapter 8). With the aid of computer techniques, it is possible to calculate the dose distribution around the implant using either the dose tables of Krishnaswamy (1979) or the specific dose rate factor for iodine-125 seeds derived by Anderson *et al* (1981a)(see §3.6). It is known that these sources have marked anisotropy and the effects of this have been studied by Ling *et al* (1979), who have produced isodose contours around cubic arrays of seeds on a

1 cm spacing between centres. From this work the authors concluded the following.

(i) Implants should be performed with closer spacing in the seed axis direction than in the transverse direction.
(ii) Centres of end sources should not be less than 2–3 mm inside the end of the treatment region, although an increase in 5–10% in the implanted activity may be used to compensate for the end effect.
(iii) For volume implants lateral peripheral seeds should be placed at a distance inside the surface of the treatment volume equal to about half the transverse inter-seed distance.

All of the above recommendations assume that the implanted activity is proportional to the average dimension and that the spacing is as prescribed by the spacing nomograph.

Whilst anisotropy is a problem, seed visualisation makes it impossible to determine seed orientation, thus the computer program BRACHY, developed at the Memorial Hospital, treats each seed as a point source. Using the specific dose rate factor for iodine-125, an isodose distribution can be constructed around the implanted seeds. From a knowledge of isodose contours, the volume enclosed by the matched peripheral dose (MPD), which is defined as the dose for which the contour volume matches the ellipsoidal volume determined from the dimensions taken at the time of the implant, can be calculated. Using a graph which relates the dose to a given volume, the MPD can be determined by graphical interpolation. The practice of increasing the average dimension by 1 cm in each direction has tended to move the MPD away from the peripheral seeds, thereby creating a more nearly ellipsoidal shape. This approach, however, does not provide adequate information on achieving the desired tumour dose prior to the implant procedure. In an attempt to overcome this problem Rosemark *et al* (1982) carried out an analysis of 50 iodine-125 implants whose doses were determined using MPD calculations. In addition, matched doses were determined for volumes corresponding to 1/2, 1/4 and 1/8 of the implanted volume. The total activity implanted was then divided by the matched dose for the corresponding implant volume or volume fraction and the values obtained plotted against the volume. From the resulting scattergram the authors proposed a linear relation between the cumulative activity per unit dose and volume. The derived dose table is found to be in agreement with the Memorial protocol for volumes larger that 20 cm^3 and it has been suggested that this linear relationship can be used for the pre-planning of iodine-125 implants.

Other workers have produced dose tables for iodine-125 volume implants which can be used in planning treatment. Krishnaswamy (1979) derived tables which give the MPD together with the maximum central dose for different cubic volumes implanted. Since, however, many lesions are spherical or ellipsoidal, these dose tables do not accurately express the true peripheral dose of the implant. To overcome some of these problems Wu *et al* (1985) proposed new calculation procedure which defined the point of dose specification to be the face-centred lattice point (or points) lying closest to the point of intersection of the longitudinal axes of the seeds passing through the geometric centre of the treatment volume and the volume boundary. Following the Paterson–Parker approach, the authors define the net minimum dose to be 10% above the absolute minimum dose determined at all surface centres of the lattice configuration of seeds enclosed in a spherical volume. In their study they calculated the activity per unit dose for various spherical and ellipsoidal volumes and showed that the activity required to deliver unit dose is not significantly determined by the spacing between the seeds, but that the shape of the implant volume does play an important factor in determining the total implant activity. This conclusion constitutes an important distinction between the net minimum dose and average peripheral dose calculated by Waterman and Strubler(1983) who showed that the average peripheral dose, which is approximately 80% of the average tumour dose, is nearly independent of the shape of the implant.

By considering the volume to be implanted and the variation in seed activities, Wu *et al* (1985) produced tables of the activity per unit dose for various implant volumes and the seed strength required to give a net minimum dose of 10 Gy to implants of varying volumes. Using these tables it is possible for the radiotherapist to prescribe treatment in terms of tumour dose rather than in the activity to be implanted.

6.5 Neutron Therapy

6.5.1 Clinical techniques

As indicated in §1.3, the use of neturons in brachytheraphy has been advocated on radiobiological grounds. Sources of californium-252 became available for clinical use in the United States in late 1968 and clinical trails began, notably at the Memorial Hospital, New York, and the MD Anderson Hospital, Houston. Initially, neutrons were used in interstitial brachytherapy for the treatment of advance disease, mainly metastatic inguinal or cervical lymph nodes or advance vaginal or intraoral lesions (Castro *et al* 1973). Due to the high biological effectiveness of neutrons, special consideration was given to maintaining a low exposure dose to personnel, thus insertions were carried out using after-loading techniques. Many workers adopted the Pierquin techniques of using solid or flexible metal guides to allow percutaneous insertion of plastic tubes or, in the case of vaginal implants, hollow stainless steel needles. Simple single- and double-plane implants, as well as volume implants, were carried out using line sources separated by 1 to 1.5 cm with the ends crossed (Castro *et al* 1973). In this work a total dose of 6000 to 7000 rem was given. Since this dose has two components, that due to photons with an RBE denoted by RBE_γ and having a value of 1, and that due to neutrons with an RBE denoted by RBE_n in the range 7 to 7.5, the authors derived their dose from

$$\text{Total dose(rem)} = \text{Neutron dose(cGy)} \times \mathrm{RBE}_n + \text{Gamma-dose(cGy)} \times \mathrm{RBE}_\gamma$$

In 1974 californium-252 seeds, contained in nylon ribbon, became available (Permar 1976). Initially, californium-252 was encapsulated in stainless steel which had the disadvantage of not being able to effectively screen the beta-rays emitted by the sources. These sources were modified by changing the encapsulating material to 0.25 mm of platinum, the ALC-P4C seeds. Unfortunately, even with these seeds the dose at 0.5 cm due to the beta-rays was twice that due to the gamma-rays. Again, after-loading techniques were adopted (Permar 1976, Vallejo *et al* 1977), the sources being loaded into either plastic tubes placed through the tumour and secured within the volume by metallic buttons on the skin surface or into implanted 16 gauge surgical needles. To assess the effective use of neutrons Vallejo *et al* (1977) set out to replicate their conventional dose of 60 Gy in six days using iridium-192 seeds and found that similar clinical effects were achieved with a dose of 840 cGy (neutron cGy) + 740 cGy (beta + photon cGy).

From their study of ten patients the authors drew the conclusion that tumour response did not differ quantitatively from that due to the photons from iridium-192 wire. A similar conclusion was drawn by Berry *et al* (1979) and Wiernik and Young (1979) who reported that in a group of 25 patients there was no evidence to suggest a therapeutic gain when using neutrons, as expressed in increased control of the squamous cell carcinoma or less late skin damage. However, Seydel (1979) reported that in a group of six patients the neutrons from californium-252 were at least as effective compared with photons, and with no increase in the complication rate, whilst Tsuya *et al* (1979) reported good tumour control in a group of 44 patients treated with neutron brachytherapy.

6.5.2 Neutron dosimetry

Various workers (Krishnaswamy 1971, 1972b, 1974, Colvett *et al* 1972, Anderson and Ding 1975) calculated and measured the dose distribution around different californium-252 sources. These data give the total dose to tissue from both the neutrons and the gamma-rays in cGy (rads) per μg h of californium-252 and have become the basis of much clinical neutron dosimetry when used either in tabular or graphic form, or expressed as a specific dose rate factor for use in the BRACHY group of programs.

To provide guide-lines for the use of neutrons for interstitial therapy, Bloch *et al* (1972) produced dose tables and a set of rules for californium-252 implants. Beginning with the two systems extensively used for radium implants, the Paterson–Parker and the Quimby systems, the authors calculated treatment data using each system with californium-252 and then extended the work to develop a modified system to achieve a more uniform dose distribution in the treatment plane. Using the Monte Carlo technique proposed by Krishnaswamy (1972b), the dose distribution for a linear source due to the neutrons was calculated using the numerical integration of a point source method. Similarly, using the calculated differential point source distribution due to gamma-rays as the product of the number of photons per μg h emitted from the source, the absorbed dose constant and the specific absorbed dose fraction, they obtained the gamma-ray component of dose from the line source by numerical integration. These calculations were verified by measurement of the neutron and photon dose in a tissue-equivalent phantom using small tissue-equivalent and aluminium ionisation chambers for a specific californium source. These measurements showed that the calculated and measured central axis neutron and gamma-dose components agreed within 6% and 4% respectively.

For planar implants it was shown that using the Paterson–Parker system for californium-252 produced comparable uniformity to that achieved with radium for implant areas greater than 3×3 cm^2, whereas the results with the Quimby system were not as good. Bloch *et al* therefore proposed a system intermediate between the two classical systems which used uniformly loaded sources uniformly distributed in the implant plane with the ends uncrossed. This relates the average neutron dose rate to the target area—table 6.18.

Table 6.18 Dose rate table and modified distribution rules for californium-252 planar implants. (Modified data from Bloch *et al* (1972).)

Implant area (cm^2)	Target area (cm^2)	Edge weighting	Average neutron dose rate (cGy per μg h)	Dose rate ratio (neutron/gamma)
2 × 2	1 × 2	1.4	2.237	1.84
2 × 4	1 × 4	1.4	1.482	1.83
2 × 6	1 × 6	1.5	1.107	1.81
2 × 8	1 × 8	1.5	0.889	1.80
3 × 3	2 × 3	1.4	1.376	1.84
3 × 5	2 × 5	1.5	0.986	1.82
3 × 6	2 × 6	1.5	0.867	1.81
3 × 9	2 × 9	1.5	0.637	1.78
4 × 4	2.5 × 4	1.4	0.976	1.82
4 × 6	2.5 × 6	1.5	0.738	1.80
4 × 8	2.5 × 8	1.5	0.597	1.78
4 × 10	2.5 × 10	1.5	0.501	1.77
5 × 5	3 × 5	1.5	0.717	1.80
6 × 6	4 × 6	1.5	0.542	1.78
6 × 9	4 × 9	1.5	0.403	1.75
8 × 8	6 × 8	1.5	0.349	1.74
9 × 9	6.5 × 9	1.5	0.268	1.71

To achieve good dose homogeneity the following rules must be obeyed.

(i) The length of time the peripheral sources are in the patient must be weighted so that the dose at the peripheral point lying on the centre line perpendicular to the axis of the the sources is within 10% of the reference maximum dose (defined by the Quimby system). This can be achieved in after-loading techniques by removing the inner sources after a precalculated time.

(ii) An area in the treatment plane is specified to be the effective target area such that 95% of that plane receives a dose within 10% of the average dose over that area. To achieve this criterion, geometric conditions were found empirically which required the effective treatment length to be increased.

These two rules are satisfied if the conditions in table 6.19 are obeyed.

The authors also found that applying the Paterson–Parker rules for cuboid implants using californium-252 also gave excellent dose rate uniformity.

Table 6.19 Conditions required to obey rules for planar implants.

Rule (i)		Rule (ii)	
Width of implant (mm)	Edge weighting factor	Length of target area, L(mm)	Length of implant (mm)
		10–20	$L+10$
Less than 50	1.4	25	$L+15$
Greater than or equal to 50	1.5	30–60	$L+20$
		65	$L+25$

Example 6.3

Assume an area of 2×6 cm^2 is to be treated with californium-252 needles containing 2 μg each and of active length 3 cm to a neutron dose of 10 Gy. Calculate the time of insertion.

From rule (ii) above the area to be implanted is 3×6 cm^2, the implant consisting of seven parallel sources at 1 cm spacing. The dose rate for such an implant, from table 6.18, is 0.867 cGy (μg h)$^{-1}$. Therefore the dose rate due to seven sources is $7 \times 2 \times 0.867$ cGy h^{-1}. The inner sources must therefore be removed after $(1000/14) \times 0.867$ hours = 82.4 hours.

From rule (i) the outer sources are to be removed at 82.4×1.5 hours = 123.6 hours. Since the neutron to gamma dose ratio for this size of implant is 1.81, the average gamma-dose will be $1000/1.81 = 552.5$ cGy.

6.5.3 Neutron protection

The mode of interaction of neutrons with matter and the resultant higher RBE means that special consideration has to be given to staff protection in operational procedures and also to the dose other sites in the patient receive during the implant. This is particularly important for implants in the head and neck region where the dose to the lens of the eye is critical. In general, neutron shields have to be made of hydrogenous materials to absorb the neutrons, which therefore tends to make such shields bulky. Thus, for californium-252, which emits both neutrons and gamma-rays, this neutron shielding is in addition to the lead which is necessitated by the presence of the strong gamma-ray component. Typical bed shields are made of 3 to 4 half value layers of water equivalent polyester(WEP) and the eye dose to staff is reduced by using a Perspex shield at the level of the head (Castro *et al* 1973, Permar 1976). Similarly, containers used for transporting the sources from the main safe to the treatment area will be large and bulky. Such protection problems represent a significant inconvenience in the use of californium-252.

7 Intracavitary Brachytherapy

7.1 Introduction

Intracavitary brachytherapy, as the name implies, involves the insertion of radioactive sources into natural body cavities. Over the years various lesions in such sites as the bladder, anus, rectum, antrum, oesophagus, nasopharynx and auditory tube have been treated in this way. The most widespread use of this form of therapy, however, has been in the treatment of gynaecological malignancies. In the treatment of carcinoma of the uterus and uterine cervix the sources, in suitable containers, are inserted directly into the vagina and the uterus. This has the major advantage that the sources are in close contact with the tumour, thus giving the neoplasm a high dose of radiation, whilst more distant tissues within the pelvis receive comparatively less dose due to the rapid fall-off in dose in accordance with the inverse square law. This effect means that for extensive tumours intracavitary brachytherapy alone does not form a complete treatment, thus it is used in conjunction with external beam therapy given before or after the intracavitary therapy.

Early work was often empirical, with little knowledge of the biological effects of the treatment and insufficient understanding of the need to filter out beta-rays by source screenage. In the 1920s and 1930s a more scientific approach to intracavitary therapy was adopted, notably at the Radiumhemmet in Stockholm and at the Curie Foundation in Paris. From this work evolved the 'Stockholm system' (Heyman 1929, 1935) and the 'Paris system' (Regaud 1929) for the treatment of uterine cancer. In both of the methods the total amount of radium to be used was divided into two, one for placement in the uterus and the other in the lateral fornices of the vagina. However, the relative amounts of radium in the uterus and vagina at each centre differed—40 to 50 mg of radium being used in the uterine applicator and 60 to 80 mg of radium in the vaginal applicator in the Stockholm system, whilst in the Paris system approximately 30 mg of radium were used in the uterine applicator and 30 mg of radium in the vaginal applicator.

The uterine sources in both systems were arranged in a line within an applicator, extending from the external os to near the top of the uterine cavity. The arrangement of the vaginal sources, however, differed. In the original Paris method the vaginal sources were placed in cylindrical 'corks' of approximately 20 mm diameter and held in place in the lateral fornices by a metal spring, whilst in the Stockholm technique heavily shielded silver or lead boxes were used to contain the vaginal sources. Another major difference between the systems was the treatment schedules. The Paris method used a continuous application of the sources lasting approximately 120 hours, whilst the radiation was fractionated into three equal applications of between 20 to 24 hours in the Stockholm technique, with an interval of one week between the first and second application and a two to three week interval between the second and third application.

Various adaptations and modifications have been made to these original techniques over the years. In the 1930s Tod and Meredith (1938), working at the Christie Hospital, Manchester, studied the Paris system and adapted it to establish what has become the basis of probably the most universally accepted brachytherapy technique. The treatment regime developed, whilst following the Paris system for relative source loadings between the uterus and the vagina, adopted a fractionated approach to treatment by using two insertions of approximately 57 hours at an interval of one week. Various techniques have also been developed for the treatment of cancer of the uterine body, which use either larger uterine applicators (which take line sources) or special applicators, the most famous of which are the Heyman's capsules (Heyman 1935, Heyman *et al* 1941). In this latter technique the uterine cavity is packed with multiple small radioactive sources, a technique particularly suited to large irregular cavities.

Changes in the choice of radionuclide and in the design of the applicators and colpostats have also occurred with time. Basically, all of the more recent systems are a modification of the Paris technique, but with the irradiation fractionated into at least two applications following the Stockholm approach. In the early 1950s Fletcher *et al* (1952), working at the MD Anderson Hospital, Houston, developed improved applicators which incorporated bladder and rectal shields. This system was later adapted to allow the sources to be placed in position after the applicator had been satisfactorily inserted in the uterus and vagina.

These after-loading techniques were first introduced by Henschke (1960) and became the forerunner of various manual after-loading techniques now available and also the commercially available treatment units which enable sources to be loaded remotely into the applicators, thereby reducing the radiation to staff. Such techniques have, in part, been made possible by the use of cobalt-60 and caesium-137 which have now replaced radium as the radionuclide of choice for intracavitary brachytherapy. Care must, however, be taken in taking radium data directly and using it with the other radionuclides because of the differences in dose distribution around the sources due to different wall filtration.

7.2 Dose Specification

In the Stockholm and Paris systems the treatment was specified in terms of milligram–hours. When intracavitary therapy is used in conjunction with external beam therapy this method of dose specification is not adequate to describe the overall treatment and other approaches have therefore been developed which specify the intracavitary therapy in terms of the absorbed dose received by tissue.

7.2.1 Manchester approach

In order to define the dose given to a lesion in a more meaningful way than mg h of radium, Tod and Meredith (1938) began to calculate the dose in roentgen to various sites in the pelvis in a more scientifically rigorous way for treatment of cancer of the uterine cervix. A series of points within the pelvis were therefore defined at which the exposure dose should be stated and measured, the points chosen being anatomically comparable from patient to patient. Due to the rapid dose variations existing in such treatment, it was decided not to choose points such as the external os or the mucosa of the vaginal vault, but rather a point where the dosage is not too sensitive to small alterations in applicator position.

The work of Todd (1938) concluded that the limiting radiation dose was not the dose to the rectum or bladder, but the dose in the area where the uterine vessels cross the ureter, the 'paracervical triangle'. It was therefore decided to regard tolerance in this region as the limiting factor in the irradiation of the uterine cervix. This led to the designation of Point A for specifying the dose and the system of sources proposed was designed to deliver a constant dose rate to defined points near the cervix, irrespective of variation in the size and shape of the uterus and vagina. Initially this point was defined as being 2 cm lateral to the uterine canal and 2 cm from the mucous membrane of the lateral fornix of the vagina in the plane of the uterus. Subsequent experience showed that for practical purposes the lower end of the intra-uterine tube could be taken to be level with the lateral fornix, thus Point A is now defined to be 2 cm along the axis of the central tube from the lower end and 2 cm from it laterally (figure 7.1). A further reference point, Point B, was also defined to be 5 cm from the mid-line and 2 cm up from the mucous membrane of the lateral fornix (Meredith 1967). This point was chosen by Tod and Meredith since it not only gives the dose in the vicinity of the pelvic wall near the obturator node, but also a good indication of the lateral spread of the effective dose.

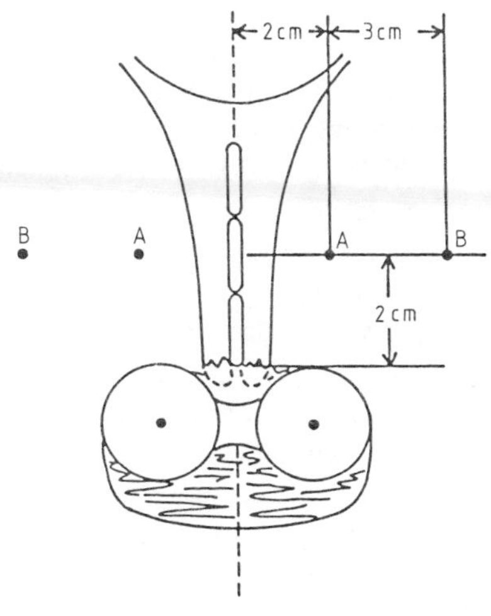

Figure 7.1 Positions of A and B for 'ideal' Manchester insertion. (Reproduced by kind permission of Amersham International plc.)

7.2.2 Houston approach

Rather than rely solely on the dose at two fixed points to define treatment, as with the Manchester system, Fletcher, working at the MD Anderson Hospital, Houston, began to consider the usefulness of adopting a three-dimensional approach to determine doses within the pelvis. One of the regions of interest in the treatment of carcinoma of the cervix are the pelvic modes and early studies by Fletcher (1953) showed that

Point B of the Manchester system did not correspond to the dose to the principal (obturator) node or hypogastric nodes. Further study by Lewis *et al* (1960) confirmed the results of Fletcher and concluded that the age of the patient appeared to influence the location of Points A and B, as did the staging of the disease.

In addition, Schwarz (1969) carried out an evaluation of the Manchester system and showed that in many studies the dose at the right Point A is not the same as the left Point A, and similarly Point B defined in the Manchester system only gave an indication of the dose to the obturator lymph nodes, since the other regional lymphatic nodes were outside the true pelvis. An extensive study of the normal lymph node anatomy of the pelvis and the para-aortic area by the routine use of lower-extremity lymphangiograms led Durrance and Fletcher (1968) to define a plane in which dose calculations should be made to achieve improved understanding of the dose around intracavitary sources.

Whilst acknowledging that it is sufficient to know the dose rate at representative points, they pointed out the danger of relying on these calculations alone and strongly contended that it is essential to know the dose gradients within the area. Initially, they suggested calculating the dose within a plane defined as being perpendicular to the mid-sagittal plane of the body, passing through a line joining the superior aspect of the pelvis and the anterior aspect of the S1–S2 junction in the sacral spine. However, it was realised that if a second plane was constructed upwards from this plane to the region of the bifurcation of the aorta, more detailed dose rate estimates could be achieved for the common iliac and lower para-aortic lymph node chains. A further plane was therefore suggested, passing through the transverse line in the pelvic brim plane to the mid-point of the anterior aspect of the body of L-4. This plane was trapezoidal in shape (figure 7.2). At the top of the trapezoid, points 2 cm lateral to the mid-line of the level of L-4 (labelled R.PARA and L.PARA) are used to estimate the dose rate to the lower para-aortic area, whilst points 6 cm lateral to the mid-line at the inferior end of the plane (labelled R.EXT and L.EXT for the right and left external respectively) give an estimate of the dose rate to the mid-external iliac nodes. The mid-point of a line (denoted by 1/2 on each side) joining these two points is used to estimate the dose to the lower common iliac lymph nodes (labelled R.COM and L.COM). To obtain an estimate of the dose rate to the obturator and hypogastric nodes the dose should be calculated at a point 5 cm lateral to the mid-line at the lower end of the trapezoid (Fletcher 1980). Using this method of defining dose, Fletcher was able to produce data of the dose given to various lymph nodes for a given number of mg h of radium.

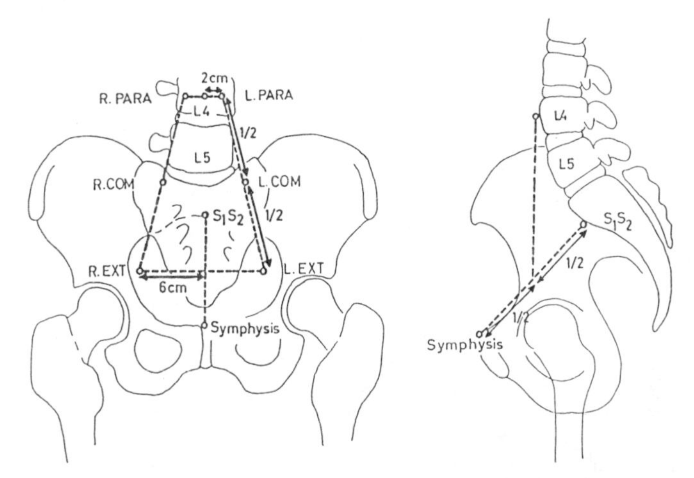

Figure 7.2 Determination of lymphatic trapezoid for the antero-posterior (AP) and lateral view. (Reproduced from ICRU (1985) by kind permission of the ICRU.)

Various attempts have been made to compare the methods of dose prescription in milligram–hours to the doses calculated at Points A and B. Cunningham *et al* (1981) carried out a comparison on 77 patients of the doses to Points A and B compared to milligram–hours prescription using a linear, least square regression technique and found a linear relationship between the dose to Points A and B and the mg h of the form

$$\text{dose in cGy to A} = (0.78\ \text{cGy mg}^{-1}\,\text{h}^{-1}) \times \text{number of mg h}$$
$$\text{dose in cGy to B} = (0.24\ \text{cGy mg}^{-1}\,\text{h}^{-1}) \times \text{number of mg h.}$$

These relationships have correlation coefficients of 0.84 and 0.88 respectively. However, Potish *et al* (1982), who carried out a similar comparison on 91 patients, concluded that even though moderately high correlation coefficients were found, the direct comparison, particularly between individual patients, was fraught with dose uncertainties of clinical significance. The conclusion, therefore, of this work was that it was not possible to directly compare the Manchester and Fletcher system of dosage and thus prescriptions from one system could not be interchanged with prescriptions from the other system.

7.2.3 ICRU approach

In 1978 the ICRU published Report 29 on 'Dose Specification for Reporting External Beam Therapy with Photons and Electrons' (ICRU 1978). In this report the concepts of target volume, treatment volume and irradiated volume were defined for external beam therapy. Since intracavitary therapy is often used in conjunction with external beam therapy, the ICRU (1985) in Report 38 on 'Dose and Volume Specification for Reporting Intracavitary Therapy in Gynaecology' have advocated that similar terminology should be used for intracavitary work with the necessary modifications in the way in which the various volumes are defined. In this report these volumes are defined as follows.

> **Target Volume:** is that volume which contains those tissues that are to be irradiated to a specified absorbed dose according to a specified time–dose pattern. For certain clinical situations there may be more than one Target Volume since it may not be always possible to treat the total extent of the disease by intracavitary therapy alone.
> **Treatment Volume:** is the volume enclosed by a relevant isodose surface which encompasses at least the Target Volume. In intracavitary therapy, the dose cannot be specified in terms of a percentage of the dose at any point within the Target Volume due to the steep dose gradients. It is therefore necessary to firstly agree upon a reference dose level, secondly to plan the treatment to include the Target Volume within that reference isodose surface and then to determine the dimensions of the Treatment Volume.
> **Irradiated Volume:** is that volume, larger than the Treatment Volume which receives an absorbed dose considered to be significant in relation to tissue tolerance.

Using these definitions, the ICRU recommend that for a gynaecological insertion of uterine and vaginal sources the specification of an intracavitary application should be in terms of a treatment volume which is the volume enclosed by the 60 Gy reference isodose surface, the dimensions of the volumes being defined by the three dimensions, measured in cm, shown in figure 7.3. These are as follows.

(i) The height (d_h) is the maximum dimension along the intra-uterine source and is measured in the oblique frontal plane containing the intra-uterine source.
(ii) The width (d_w) is the maximum dimension perpendicular to the intra-uterine source and is measured in the same oblique frontal plane.
(iii) The thickness (d_t) is the maximum dimension perpendicular to the intra-uterine source and is measured in the oblique sagittal plane containing the intra-uterine source.

In addition to these parameters it is also recommended that the

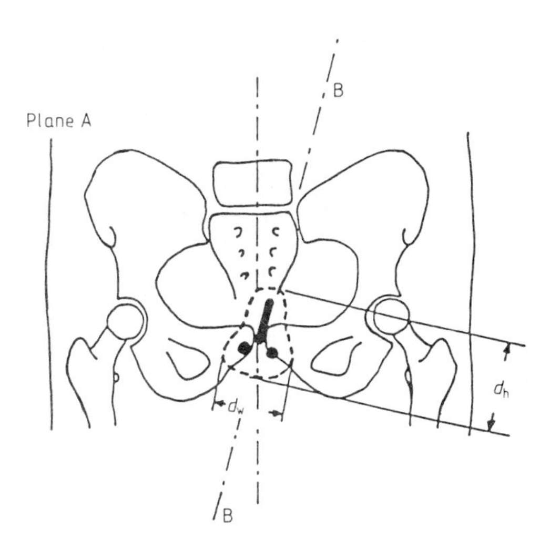

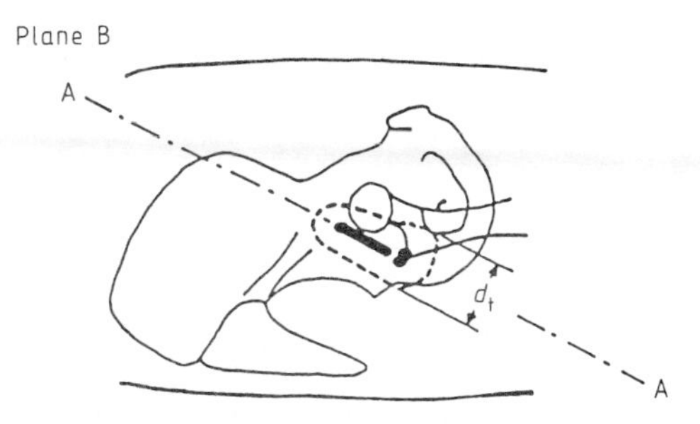

Figure 7.3 Geometry for measurement of the size of the pear-shaped 60 Gy isodose surface for typical treatment of cervix carcinoma using a central uterine tube and two vaginal applicators. Plane A is the 'oblique' frontal plane containing the central tube around a transverse axis. Plane B is the 'oblique' sagittal plane that contains the intra-uterine applicator. (Reproduced from ICRU (1985) by kind permission of the ICRU.)

absorbed dose at reference points related to bony structures, such as the lymphatic trapezoid and pelvic wall, should be reported. The lymphatic trapezoid is obtained in a similar manner to that described for the Houston approach, whilst the pelvic wall reference point, intended to be representative of the absorbed dose at the distal part of the parametrium and at the obturator lymph node, can be visualised on an antero-posterior (AP) and a lateral radiograph and related to bony structures. On the AP film the pelvic wall reference point is defined as being the point intersected by two lines; a horizontal line tangential to the highest point of the acetabulum and a vertical line tangential to the inner aspect of the acetabulum (figure 7.4). On the lateral film the highest points of the left and right acetabulum, in the cranio-caudal direction, are joined and the lateral projection of the pelvic wall reference point is located at the mid-distance of these points. Defining these points is particularly important when combining intracavitary and external beam therapy.

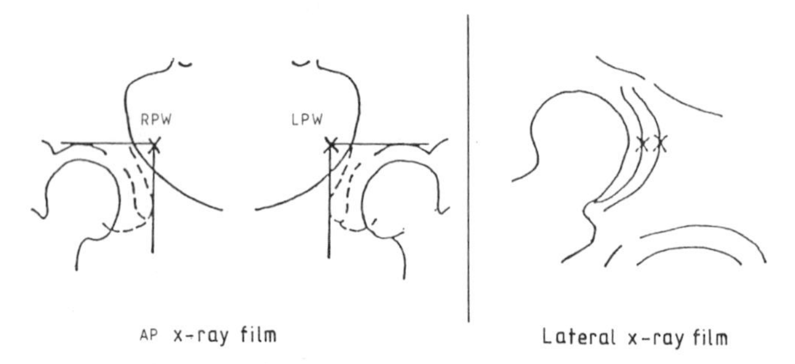

Figure 7.4 Determination of the right (RPW) and left (LPW) pelvic wall reference points. (Reproduced from ICRU (1985) by kind permission of the ICRU.)

Due to the steep dose gradients in the region of the sources, specifying a treatment in terms of dose at a point is not recommended. It is therefore proposed that the total intracavitary treatment should be specified in terms of total air kerma, which is the sum of the product of the reference air kerma rate and the duration of the application of each source and is proportional to the number of mg h. The 'reference air kerma' for a source is defined as the air kerma rate to air, in air, at a reference distance of 1 m, corrected for air attenuation and scattering. These recommendations follow those made by the CFMRI (Comité Français pour la Mesure des Rayonnements Ionisants) (1983). The use of the reference air kerma term is shown in Example 7.1. It is also recognised that to completely specify the treatment the time–dose pattern should also be given.

Example 7.1

In a gynaecological treatment 80 mg of radium are inserted for 50 hours. What is the reference air kerma rate, and what will be the total air kerma if tubes filtered with 1 mm Pt are used?

A point source containing 1 mg of radium (0.5 mm Pt filtration) produces an exposure rate of 0.825 $\mathrm{mR\,h^{-1}}$ at 1 m. Since the conversion factor from exposure to air kerma is 0.873 $\mathrm{cGy\,R^{-1}}$ and the correction for the difference in filtration is 0.93, then a 1 mg tube of 1 mm Pt filtration produces an air kerma rate of $(0.825 \times 0.873 \times 0.93) = 6.7\ \mu\mathrm{Gy\,h^{-1}}$ at 1 m. Therefore, for the total of 80 mg of radium used, the reference air kerma rate is $(6.7 \times 80) = 536\ \mu\mathrm{Gy\,h^{-1}}$ at 1 m. Since the time of the insertion is 50 hours then the total air kerma is $536 \times 50 = 2.7$ cGy at 1 m.

7.3 Manual Gynaecological Treatment Systems

7.3.1 Manchester system

Making the basic assumption that the uterus lies in the mid-line of the body, Tod and Meredith (1953) produced loading rules for the uterine and vaginal sources to enable the same exposure to be given to Point A regardless of the shape and size of the uterus and vagina, and consequently the applicators used. The uterine tubes which are made of thin moulded rubber or plastic, closed at one end and having a flange at the other to aid fixation, are available in three lengths and can contain either one, two or three 20 mm long radium/caesium tubes. The vaginal applicators developed for this system, ovoids, are a modification of the cork colpostats of the Paris technique and are made of hard rubber or plastic and have diameters of 20, 25 and 30 mm (figure 7.5). The shapes of the ovoids are approximately ellipsoids of revolution, this being the shape on the isodose surface around a radium tube of 15 mm active length and more recently 13.5 mm active length. In use, the ovoids are employed in pairs, one ovoid being placed in each of the lateral fornices at the level of the cervix and fixed in position by the use of a 'spacer' separating the ovoids by 10 mm or by a 'washer' which allows them to lie almost in contact with each other. In order to reduce the rectal dose, gauze to a thickness of at least 15 mm is packed behind the ovoids. This packing also serves to prevent ovoid movement during treatment (Cole 1963).

With the wide choice of applicators to suit the individual anatomy, the arrangement of sources was defined so as to ensure the dose to Point A would be the same for all insertions, whilst at the same time ensuring that the tolerance to the vaginal mucosa was not exceeded. To achieve a constant dose rate at Point A, the Manchester system defined the number of 'units' of radium to be used in each ovoid and tube size. The relative

Figure 7.5 Manchester system applicators. Pairs of ovoids (small, medium and large), intra-uterine tubes (long, medium and short), together with washer (bottom left) and spacer (bottom centre). (Reproduced by kind permission of Amersham International plc.)

units, given in table 7.1, were based on experience which showed that the vaginal sources should not contribute more than about 1/3 of the total dose at A, a similar ratio to that found by Kottmeier (1951) following a study of a series of cases treated at Stockholm.

Table 7.1 Arrangement of sources to achieve a constant dose rate at Point A.

Application	Number of tubes	Units
Intra-uterine tube		
Long applicator	3	4–4–6
Medium applicator	2	4–6
Short applicator	1	8
Vaginal ovoids		
Large	1 (ideally)	9 units each
Medium	1 (ideally)	8 units each
Small	1 (ideally)	7 units each

The size of the unit of radium was initially determined from experiences of tolerance which indicated an exposure dose of 8000 R in 144 hours total to Point A, the treatment given in two equal fractions of

4000 R. With this dose, the 'unit' was defined as 2.5 mg of radium with 1 mm Pt filtration.

Certain clinical situations do not allow the standard insertion and it is only possible to use a short central tube. In this situation the dose rate at A is reduced by 13% when used in conjunction with standard ovoids. Similarly, a reduction in dose rate is necessary when it is not possible to use two ovoids side by side in the fornices. Rather than not use vaginal sources in conjunction with a central tube the two ovoids are placed in tandem, thereby reducing the dose by 7%. With currently available radium sources or radium substitutes it is not possible to exactly fulfil the vaginal ovoid loadings with the large and small ovoids. One solution proposed (Meredith and Massey 1977) is that to achieve nine units in a large ovoid, for example 22.5 mg of radium or radium-equivalent, it would be possible to use a 25 mg tube (G5) for the first insertion and a 20 mg tube (G4 type) for the second insertion, both for the same time. This, however, allows the possibility of error and thus it is recommended that the ovoids be loaded with three G-type tubes, 5 + 10 + 10 mg. Whilst this places 25 mg of radium within the ovoid, the net effect of the set of sources is to reduce the dose rate to Point A by 10% due to the extra filtration of the sources. This is equivalent to an effective content activity of 22.5 mg in the ovoid. Typical dose rates to Point A from various standard loadings with washers and spacers are given in table 7.2 for the G-type tubes.

Table 7.2 Dose rates with G-type tubes.

Applicator	Loading (mg Ra eq)	Dose to Point A (cGy per hour)
Intra-uterine		
Long	10 + 10 + 15	34.4
Medium	10 + 15	34.2
Short	20	27.3
Ovoid tandems		
Large—with spacer	5 + 10 + 10	18.3
—with washer		18.9
Medium—with spacer	20	18.8
—with washer		19.0
Small—with spacer	5 + 5 + 10	19.4
—with washer		19.5

Using table 7.2, it is possible to calculate the dose to Point A by adding the contribution of the tube and the ovoid tandem. For example, for a medium central tube with a 'tandem' of medium ovoids, separated by a spacer, the dose rate is 53.0 $cGy\,h^{-1}$ (34.2 $cGy\,h^{-1}$ + 18.8 $cGy\,h^{-1}$). It is found that for standard arrangements the dose to A is within the range of 52.8 $cGy\,h^{-1}$ to an error of $\pm 1.5\%$. The dose at Point B is of the order of 25%–30% of the dose at Point A. With the advent of computers, it has been possible to produce dose distributions around typical insertions.

It must be remembered that these dose rates apply for an ideal insertion. Often the disease does not allow the ideal positioning of sources, the uterine canal being pushed away from the mid-line. In such cases, Point A is considered to be carried with the uterus and is thus at a point 20 mm from the uterine canal and perpendicular to it, whereas Point B remains in the same position relative to the patient mid-line (figure 7.6). Wilkinson *et al* (1983) pointed out, however, that even when the sources are displaced, it is the dose rate at Point A for the *ideal insertion* which determines the treatment time, even though the reference geometry is rarely achieved.

The Manchester technique for the treatment of carcinoma of the body of the uterus, where intracavitary therapy is indicated, is similar in

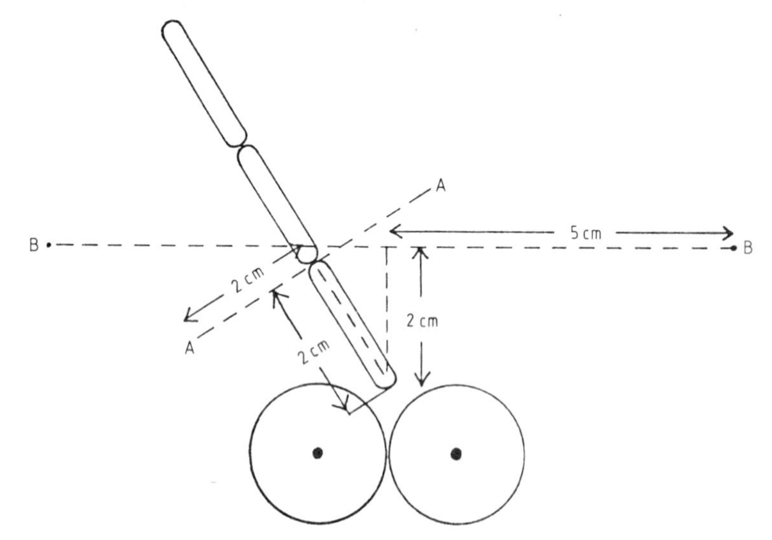

Figure 7.6 Relative position of Point A and Point B for a 'non-ideal' insertion. (Reproduced from Meredith (1967) by permission of Churchill Livingstone.)

concept to their treatment of uterine cancer, except that the sources are loaded in such a way that a greater part of the dose comes from the uterine applicator, thus the relative loadings of uterine and vaginal applications are different. In this technique the whole length of the uterus is treated up to the fundus, using a 1 cm diameter uterine applicator loaded with a greater number of sources (Paterson 1963). The amount of radium or radium-equivalent isotope and the relative loadings used vary with the length of the uterus and are given in table 7.3, where each unit is 5 mg of radium. The dosage is the same as for the uterine cervix treatment, i.e. 8000 R to A in two insertions of 72 hours, although the dose distribution will be significantly different.

Table 7.3 Loading of sources.

Uterus (Loading from fundus to cervix)	Vagina (Loading of ovoids)
5 2 2 3 2 2 units	
5 2 2 2 2 units	Large ovoids 4 units
5 2 2 2 units	Medium ovoids 3 units
5 2 1 units	

7.3.2 Stockholm system

The original Stockholm technique devised by Forsell in 1914 was based on the application of relatively large amounts of heavily filtered radium salt, the filter being equivalent to 3 mm of lead. Sources for insertion into the uterine cavity were placed in screw-ended applicators, based on Domnici tubes, the walls of which were equivalent to 2 mm of lead. The vaginal applicators, however, took the form of flat boxes made of 1.7 mm lead lined with 0.45 mm of silver, the boxes varying in size from $56 \times 36\ \text{mm}^2$ to $25 \times 16\ \text{mm}^2$ (Heyman 1929) and containing up to ten or twelve radium tubes placed in lines parallel to each other, the height of the boxes being of the order of 7.5 mm.

In clinical use the technique of insertion varies according to the shape and extent of the tumour, the selection of the applicators being chosen to ensure coverage, as far as possible, of the entire surface of the tumour and the distention of the vagina laterally to spread the irradiation as far as possible to the lateral pelvic walls. The amount of radium used is such that over the three insertions the uterine cavity receives between 2000 and 2600 mg h of radium and the vagina about 4500 mg h. A review of the Stockholm system has been given by Kottmeier (1964).

Over the years the Stockholm system has been modified, notably by Hurdon (1941) at the Marie Curie Hospital, London, who used a silver intra-uterine central tube loaded with two 25 mg radium tubes and three silver boxes for insertion into the vagina, each box containing 20 mg of radium in the form of four 5 mg small tubes. In cases of a narrow vaginal vault, two silver boxes containing 25 mg of radium were inserted. The design of the boxes is such that the filtration is equivalent to 1 mm of platinum. The clinical technique was to insert an intra-uterine tube and then to arrange one box to lie across the cervix with the other boxes placed in the lateral fornices. The boxes are maintained in position for the duration of the insertion by careful packing. This packing also ensures that the boxes do not lie against the lateral wall of the vagina. Calculations have shown that the typical dose rate at Point A is approxi-

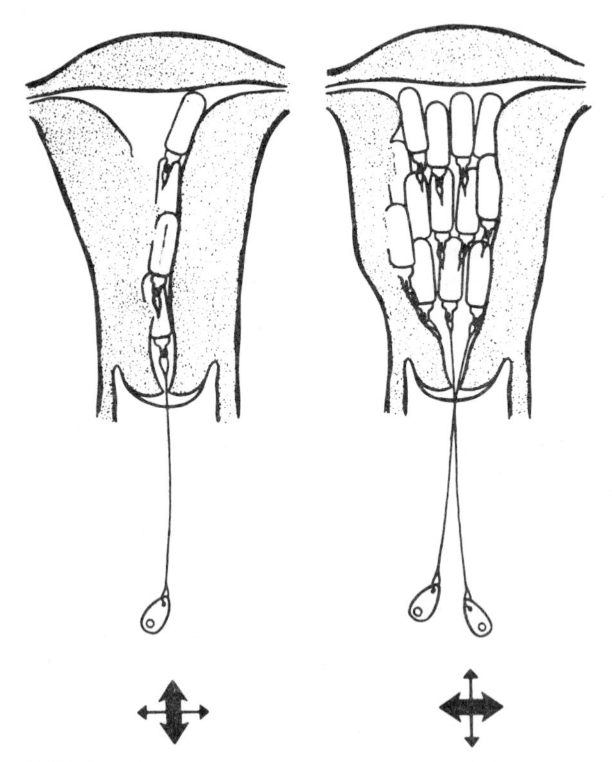

Figure 7.7 Insertion of Heyman capsules in the treatment of carcinoma of the uterus. (Reproduced from Fletcher (1980) by kind permission of Lea and Febiger.)

mately 95 cGy h^{-1}, thus a total dose of 60 Gy is achieved with three insertions of approximately 22 hours. Other modifications have also been devised which have included placing the vaginal sources within Perspex boxes. Due to the shape of the vaginal applicators, the dose variation around the box is not constant and can vary by a factor of two.

The treatment of the body of the uterus in the Stockholm technique involves packing the uterus with multiple small radium sources. These are called Heyman capsules. The capsules, which are 20 mm long and 2.8 mm outside diameter, contain 8 mg radium and are loaded into four different sized containers whose outside diameters vary from 7.8 to 13.8 mm. The Heyman packing technique aims to deliver an even dose to the whole of the myometrium by packing between 10 and 20 capsules into the uterus (figure 7.7). The technique of packing these sources necessitates using an introducer to place each of the loaded capsules into the uterine cavity by progressively stretching the uterus until no more capsules can be inserted. In the procedure, care must be taken to ensure that the uterus is not perforated. Each capsule is threaded with a retention wire which is tagged and numbered. As they are placed in position the number is recorded which enables the sources to be removed in reverse order to insertion, thereby avoiding problems when the sources are removed. A dose of 60 Gy is prescribed to a point 15 mm from the nearest container (Kottmeier 1959), two applications being given with a three-week interval between them.

7.3.3 Fletcher applicators

In the early 1950s, Fletcher, at the MD Anderson Hospital and Tumor Institute, Houston, developed preloadable colpostats. These colpostats were designed so that they could be easily applied and maintained in the lateral fornices, allowing an increase in paracervical and parametrial irradiation by better radium distribution. In addition, shielding was included within the colpostats to provide protection to the base of the bladder and the anterior wall of the rectum, tissues which are susceptible to radiation damage and have a tolerance dose of 60 Gy and 50 Gy respectively (Fletcher 1953). These Fletcher applicators were designed along the lines of the Manchester system to take advantage of the size of the vagina, to increase the milligram–hours and therefore the lateral dose. Depending on the diameter of the ovoids and the source strength used in the uterine tube, the number of milligram–hours varied from 7000 to 12000 mg h without undue radiation reaction in the vaginal vault and the cervix, and without the risk of exceeding the tolerance dose of the rectum and bladder. In an attempt to eliminate some of these problems, the ovoids were placed in the fornices on separate handles which could be locked by a scissor joint. This had the effect of positioning the ovoids correctly and ensuring that they remained in position during treatment without the possibility of rotation or slipping during packing.

The Fletcher applicators have the same diameters as the Manchester ovoids, but are slightly longer. They are not oval in shape and hence do not follow an isodose surface. To achieve the variation in size plastic jackets were made to make the two larger size ovoids, 25 and 30 mm in diameter, and the separation achievable between the vaginal sources was in the range 20 to 60 mm for radium sources (figure 7.8(*a*)). In order to reduce the dosage in the direction of the rectum and bladder, the tungsten insert at each end was shaped so that maximum screenage is provided on the anterior rectal wall and the tongue of the bladder without decreasing the radiation to the uterosacral and broad ligaments (figure 7.8(*b*)). The loading of the applicators depended on the separation, being 15 to 20 mg, 20 to 25 mg and 25 to 30 mg of radium respectively in the small, medium and large cylinders. The uterine tube is loaded with 15 + 15 + 10 mg or 15 + 10 + 10 mg, depending upon its length. Using this technique, a maximum of 10 000 mg h is given in two insertions of 72 hours each, separated by a period of ten days. This upper limit of milligram–hours is only used for tumour sizes of less than 1 cm and is reduced for larger tumour sizes (Fletcher 1980).

Direct measurements were made of the rectal and bladder dose using a scintillation counter probe and found to be as high as 7500 R for the bladder and 7000 R for the rectum using the standard insertion times. With this technique the dose to the vaginal vault was of the order of 12 000 R to 20 000 R, compared with up to 40 000 R by other techniques not based on the Manchester/Fletcher system. It has been emphasised by Delclos *et al* (1978) that the tables outlining milligram–hours and application times can only be used with the Fletcher applicator, since without the correct shielding severe reactions and complications can occur within the patient. In addition, care must also be taken when applying the Fletcher dosage system using other radionuclides such as caesium, since the dose distribution will be affected by the different altenuations of the caesium gamma-rays within the shielding compared to the gamma-rays from radium.

For the treatment of carcinoma of the body of the uterus, the MD Anderson technique is to use Heyman capsules in conjunction with the Fletcher applicator system. Clinically, such treatments are often given

(*a*)

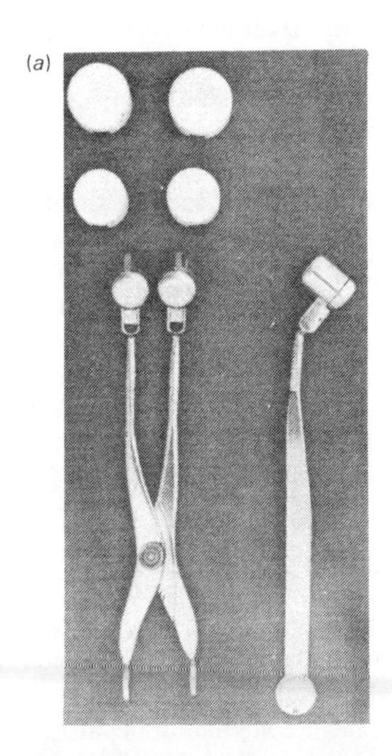

(*b*)

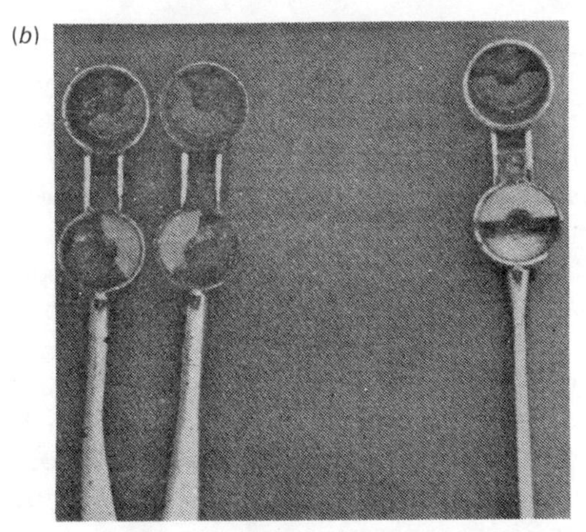

Figure 7.8 (*a*) Original Fletcher preloadable colpostats with three cylinder sizes. (*b*) Original Fletcher preloadable colpostats showing the positioning of the tungsten shielding. (Reproduced from Delclos (1984) by kind permission of Lea and Febiger.)

prior to a hysterectomy, the treatment regime being two insertions of 2500 mg h, three weeks apart (Fletcher 1980).

7.3.4 Other intracavitary systems

The development of other manual systems for the treatment of carcinoma of the uterine cervix and the body of the uterus are adaptations of one or other of the above techniques. Ernst (1949) developed an applicator system which has been widely used in the US, whilst Blomfield (1961) described a technique using a vaginal applicator which consisted of a kidney-shaped source container, with Perspex spacers mounted on a metal tube. To account for variations in individual anatomy, various sizes of the kidney-shaped holders are available and in each tungsten or brass inserts can be used to reduce the dose to the rectum. In this technique no attempt is made to fix the uterine source and the vaginal applicator together. It is possible to use more than one kidney-shaped container on the vaginal applicator in order to treat tumour extension down the vaginal wall. Prior to inserting the loaded applicator, a dummy applicator is used to gauge the size of the applicator required for treatment. Once the loaded vaginal applicator is inserted it is held in place by an abdominal belt. This technique has the advantage that no gauze packing is needed to ensure that the vaginal applicators do not move during treatment.

The introduction of megavoltage therapy equipment for external beam radiotherapy led certain centres to consider the possibility of using a single line source for the vaginal applicator, rather than using vaginal colpostats. Campbell and Douglas (1966) described the use of such a system which involved using an intra-uterine applicator of stainless steel loaded with either two or three G-type radium tubes and a vaginal applicator which consisted of a stainless steel tube capable of holding radium sources. Onto these tubes are threaded Perspex bobbins, each of which incorporated a 7 mm thick lead shield to reduce rectal dose. To avoid the possibility of these bobbins rotating, thereby negating the effect of the rectal shielding, a notch was made in each bobbin to engage with a ridge on the stainless steel tube. In the majority of these patients two bobbins were used and the remaining portion of the steel tube is fitted with smaller Perspex spacers. The 25 mg G-5 radium tubes used in the vaginal applicator are loaded into a chromium-plated brass tube, the number of tubes required corresponding to the number of bobbins used and each source being positioned so that it lay exactly opposite the lead screen of the radium source. The dosage regime used was based on the dose to the vaginal mucosa and this was set at an average of 10 000 R, although it was recognised that due to the shape of the applicator the actual dose on the vaginal wall could vary between 10 000 R and 13 000 R, and 6000 R on the posterior aspects shielded by the lead inserts.

With this type of applicator, isodose patterns are not pear-shaped but approximate to a cylinder. This has the possible advantage of making the matching of external beam therapy and intracavitary treatment easier. Neary (1943) emphasised that the difference in pelvic wall dose from a central line source and vaginal colpostats was within 5% and thus the need for external irradiation to give a cancericidal dose to the pelvic wall nodes was essential. Following this reasoning, Glazebrook (1974) suggested that the use of colpostats as a means of raising the pelvic wall dose was perhaps obsolete.

Other techniques for treating carcinoma of the body of the uterus have included the use of a radioactive cobalt spring (Strickland 1971). The use of this type of applicator was developed to try to overcome the high risk of perforation using a 'packing' technique. The applicator developed consisted of a closely wound helix of stainless steel wire formed into a loop, the ends of which are pushed into holes drilled into a stainless steel cylinder. Cobalt-60 sources, 2 mm in diameter and of activity 6.5 mCi, are then loaded into the loop, a steel pellet being inserted between each cobalt source to ensure even distribution around the loop except at the fundus of the uterus where six sources are placed together. Using this technique a total dose of 60 Gy in two insertions of about fifteen hours each is obtained to a point 15 mm from the spring. Another technique is the Marie Curie Hospital method of placing two small 8 mg containers into the upper corners of the uterine cavity and a central tube containing $8 + 25 + 25$ mg of radium. Two flat boxes, each holding 4×5 mg, are placed in the lateral fornices and a third box can be placed over the external os of the growth involving the cervix.

7.3.5 Vaginal applicators

For lesions of the vagina, applicators have been designed which consist of a series of gynae tubes loaded centrally with a Perspex cylinder, the number of sources required being dependent on the length and site of the vaginal tumour. Since the applicators have varying outside diameter, the strength of the sources within the applicator needs to be adjusted to achieve the same dose on the applicator surface for all applicators.

One method of determining the dose on the applicator surface and at points in tissue at specified distances from the centre of the active length of the applicator is to make use of table 7.4, which has been derived from the work of Paterson and Parker in the 1930s by using the correction factor of 1.08 as suggested by Gibbs and Massey (1980). When the treatment length requires a number of sources placed in a line, Parker stated that the tables could be used provided the distance between adjacent active ends should not exceed $h/2$, although with an odd number of tubes a spacing of h is just permissible (Paterson and Parker 1934). These tables can also be used to determine the percentage depth dose beyond the surface of the applicator, as mentioned in previous chapters.

Example 7.2

A vaginal lesion is to be treated with a 3 cm diameter applicator containing three 10 mg G-type radium tubes. Calculate the dose rate on the surface of the applicator at the mid-point of the sources and the dose rate at a distance of 0.5 cm.

For a G-type tube: total length = 20 mm; active length = 13.5 mm. Therefore, effective active length = (60 − 6.5) mm = 53.5 mm. From table 7.4 the number of mg h for 10 Gy on the surface ($h = 1.5$ cm) = 569. Dose rate due to 30 mg of radium = $(1000 \times 30/569)$ cGy h^{-1} = 52.7 cGy h^{-1}. Similarly, the dose rate at 0.5 cm from the applicator is given by $1000 \times 30/856 = 35.1$ cGy h^{-1}.

For Example 7.2 it must be remembered that the dose variation along the applicator surface will not be even. From the data given in Chapter 3 we know that the dose variation along a line parallel to the source axis is a function of the distance from the source. To achieve more accurate detail of dose use must be made of tables which consider the effects of this dose variation, both in terms of distance and end corrections due to obliquity. Methods of calculating dose along the length of a vaginal applicator can therefore be made using isodose distributions around sealed sources or by using the appropriate dose rate tables; for example the Young and Batho tables or the Shalek and Stovall tables for radium, or the Breitman tables for caesium tubes. If the Shalek and Stovall tables are used to calculate the dose rate to the surface of the applicator the value obtained is 51.9 cGy h^{-1}.

Example 7.3

Consider an applicator of diameter of 2.5 cm loaded with three 10 mg radium-equivalent G-type caesium tubes, each tube being 20 mm in

Table 7.4 Number of mg h required for 10 Gy for line source.

Active length	Distance from source centre						
	0.50	0.75	1.00	1.50	2.00	2.50	3.00
	(0.5 mm Pt filtration)						
0.0	32	71	129	289	514	804	1157
0.5	36	76	131	294	516	807	1159
1.0	41	83	137	299	523	813	1169
1.5	51	93	149	310	537	825	1183
2.0	59	106	165	325	554	842	1203
2.5	69	121	181	346	578	864	1228
3.0	80	135	199	370	603	890	1256
3.5	91	151	219	397	630	921	1287
4.0	102	165	238	423	660	957	1320
4.5	113	183	258	450	693	995	1358
5.0	125	198	280	478	729	1034	1398
6.0	149	231	321	538	801	1118	1488
7.0	172	264	364	597	877	1207	1588
8.0	194	296	410	661	958	1300	1693
9.0	216	332	455	725	1040	1395	1803
10.0	239	364	500	789	1125	1495	1916
12.0	284	433	590	921	1292	1699	2154
14.0	330	503	681	1054	1463	1911	2398
16.0	377	575	772	1188	1639	2128	2648
18.0	426	644	866	1322	1818	2344	2902
20.0	472	715	958	1460	1997	2561	3164
	(1.0 mm Pt filtration)						
0.0	36	80	143	322	570	891	1283
0.5	42	84	146	326	576	895	1286
1.0	51	92	154	335	584	903	1296
1.5	59	105	168	346	598	917	1312
2.0	70	121	186	367	618	936	1333
2.5	81	138	205	390	643	964	1361
3.0	94	154	228	417	672	998	1393
3.5	108	174	252	448	706	1035	1430
4.0	121	191	274	478	743	1075	1472
4.5	134	210	299	512	781	1117	1517
5.0	148	230	324	544	824	1163	1568
6.0	175	269	375	616	912	1260	1676
7.0	202	309	427	689	1004	1367	1795
8.0	230	351	480	766	1101	1481	1918
9.0	257	393	535	842	1199	1597	2046
10.0	286	434	590	921	1296	1716	2182
12.0	342	518	699	1081	1501	1961	2468
14.0	400	603	810	1243	1709	2214	2765
16.0	456	688	922	1406	1922	2472	3067
18.0	513	773	1035	1572	2136	2732	3376
20.0	570	859	1145	1739	2552	2998	3688

length. Using data from the Breitman tables for a G-type tube (given in table 7.5 below), the dose rate at 1 cm intervals along the tube is given in table 7.6, there being no spacers between the sources.

Table 7.5 Data for a 10 mg Ra eq caesium G-type tube.

Distance along source axis (mm)	Dose rate (cGy h^{-1}) for specified distances (mm) of cross plot lines from centre of source						
	0	10	20	30	40	50	60
0	—	69.1	18.8	8.4	4.7	3.0	2.0
10	104.0	40.6	15.3	7.6	4.4	2.9	2.0
15	31.4	24.3	12.3	6.7	4.1	2.7	1.9
20	15.8	14.8	9.5	5.8	3.7	2.6	1.8

Table 7.6 Dose rates along vaginal applicators loaded with three 10 mg Ra eq caesium G-type tubes.

Position relative to applicator	Dose rate (cGy h^{-1}) at specified intervals (mm) along applicator						
	−30	−20	−10	Centre	10	20	30
Surface of applicator	40.6	65.2	70.7	75.2	70.7	65.2	40.6
5 mm from surface	27.0	38.9	44.0	46.0	44.0	38.9	27.0

It is apparent that the dose increases towards the centre of the vaginal applicator. In order to even out this variation the central tube should have a lower activity. The amount of activity which should be placed in the central source has been expressed by Johns and Cunningham (1983) as a percentage of the total. However, the sources available do not always satisfy this relationship, thus a compromise has to be accepted.

Various designs of vaginal applicators and source loadings have been made to minimise the dose variation along the applicator. Dobbie (1953) designed an applicator which attempted to make the shape correspond to an isodose curve. To achieve this it is sometimes necessary to place the source at the upper end of the applicator at right angles to the axis of the other sources (Rubin *et al* (1963). Further sets of vaginal applicators have been designed by Murphy (1967), Delclos *et al* (1970), Wang (1975) and Biggs *et al* (1986).

7.4 Gynaecological After-loading Techniques

7.4.1 Advantages and disadvantages

It was not until 40 years after Abbe performed the first interstitial thereapy after-loading technique that after-loading techniques began to gain popularity for intracavitary brachytherapy. After-loading is the name given to techniques in which applicators are positioned in a patient and the radioactive sources introduced at a later stage. Early development in intracavitary after-loading techniques was carried out by many workers, including Tudway (1953) using an iridium applicator, Henschke (1960), Ridings (1963) and Suit *et al* (1963) using radium, and Horwitz *et al* (1964) who advocated the use of caesium-137 rather than radium as an after-loading technique for treating carcinoma of the cervix.

After-loading has the advantage that time can be taken to place the applicators in the optimum position in the cavity without sources being in the applicator with the consequent irradiation of staff. In traditional intracavitary treatments of cervical cancer, three main groups of staff are subjected to irradiation. These are as follows.

(i) The technical staff involved in loading sources into ovoids and intra-uterine tubes prior to use in theatre and the subsequent cleaning of the applicators on removal from the patient.
(ii) Medical staff who perform the insertion and the attending theatre nursing staff, together with the radiographers who carry out the radiographic localisation of the sources.
(iii) The staff on the wards who nurse the patient during the insertion and who are involved in the removal of the sources.

The reduction in exposure dose to staff due to after-loading techniques has been documented by many workers, including Suit *et al* (1963), Haybittle and Mitchell (1975) and Redpath *et al* (1976), the amount of reduction being dependent upon the degree of after-loading practised. In essence, after-loading techniques can be divided into two main categories, manual and remote after-loading. Manual after-loading techniques allow time for the applicator to be placed into the correct position and then the sources are manually loaded into the applicators once the patient has returned to the ward after the operation and radiographic procedures have been completed. This approach minimises the exposure to the staff in points (i) and (ii) above, but it does little to reduce exposure to ward staff. Recently, remote after-loading systems have been introduced which reduce the exposure to all staff since the sources are remotely loaded into the applicators and are removed automatically when the patient requires nursing care. Table 7.7 summarises the staff who are subjected to radiation in the various methods used in treatment techniques.

Table 7.7 Staff at risk due to exposure to radiation in brachytherapy†.

Staff	Conventional	Manual after-loading	Remote after-loading
Medical	×	0	0
Theatre nursing	×	0	0
Ward nursing	×	×	0
Radiographic	×	0	0
Technical	×	0	0

† A zero denotes no risk and a cross denotes a risk.

The cost of the after-loading options available varies markedly with the degree of after-loading required. In conventional techniques and certain manual after-loading techniques the sources used are conventional gynae radium tubes or their caesium equivalent. Other manual after-loading techniques and remote after-loading systems make use of the smaller caesium or cobalt sources. These sources are greater in cost and, depending on the equipment used, a wide range of sources or source trains may be required. In addition to these costs, the equipment itself is expensive. With high-dose-rate systems it is also necessary to have specially designed treatment rooms with the consequent building costs. The change, therefore, from conventional manual techniques to after-loading systems has major cost implications which have to be balanced against the reduction in exposure to staff (this can be as much as a factor of five to ten) and also the improved positioning of the sources within the patient. The various techniques available are discussed below.

7.4.2 Manual after-loading

The early manual after-loading techniques were a direct development of the conventional preloaded applicator systems, often using the existing radium sources. Henschke (1960) described an after-loading applicator which allowed sources loaded in thin tubing to be fitted into the three larger tubes which formed part of the applicator. In this technique the uterine applicator was first inserted, followed by two ball-shaped colpostats with empty tubes attached, one into each of the right and left lateral vaginal fornices. These lateral colpostats were fixed in position with an alignment plate and the inserted applicator was maintained in position by additional balls located on the central tube, together with the customary packing.

Other workers adopted a different approach by modifying existing preloaded applicators, whilst at the same time essentially maintaining the dosimetry associated with them. Suit *et al* (1963) described a modification to the Fletcher ovoid system which allowed the applicators to be used in an after-loading manner with standard size radium tubes. In order to accommodate these sources the colpostats had to be 1 mm greater in length whilst maintaining the same diameter, although it was necessary to rotate the shielding medially to overcome some of the mechanical difficulties encountered with the radium carrier and the rectangular shape of the handle. In this system a double hinge mechanism was required and attached to the side of the radium carrier.

Further improvements to the Fletcher/Suit applicators were made during the 1960s by Green *et al* (1969) who replaced the rectangular handles by less bulky round handles and introduced the use of a rubber plunger to push the radium source through the tube into the ovoid. The tube drops into a slot in the ovoid by gravity and is held in position by the pressure of the rubber plunger and the tube capping. Retrieval of the source at the end of treatment is achieved by tying a suture to the tube eyelet. These various systems, developed at the MD Anderson Hospital, Houston, are shown in figures 7.9 (*a*) and (*b*).

These applicators have also been used with caesium-137 tubes, but it has been shown by Saylor and Dillard (1976) that the calculated dose contribution from the ovoid sources can be in error by as much as 25% unless correction is made for the different absorption of the caesium gamma-ray photons in the applicator. Delclos *et al* (1978) have also advised extreme caution in the use of the Fletcher/Suit applicator, and that adequate checks must be made on the applicators to ensure that their construction exactly follows that developed by the originators of the applicator. Various commercially manufactured applicators were studied by Delclos and it became apparent that even though the outside appearance and measurements were correct the radioactive sources could be off-centre, the recommended tungsten shielding was not in position and the ovoid wall thickness was not always the same. These deficiencies in manufacture could result in a five-fold increase in the ovoid surface dose and the dose to the bladder and rectum could be increased by 15 to 20%. The need for caution was also emphasised by Delclos when the tables

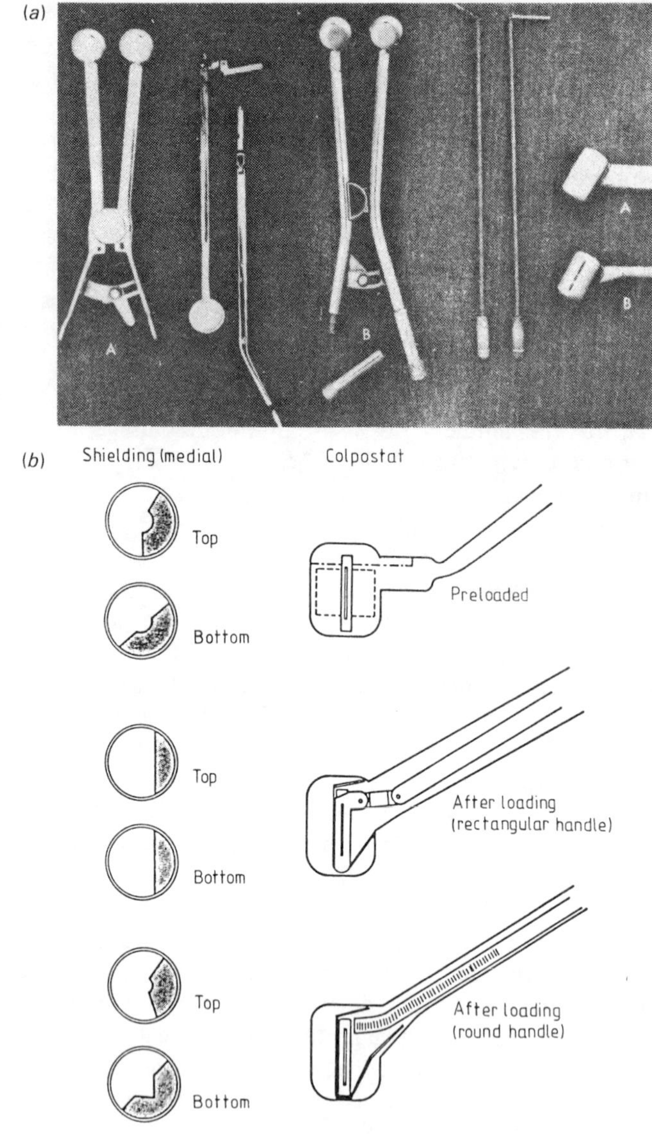

Figure 7.9 (*a*) Fletcher after-loading colpostats: A, Fletcher/Suit rectangular handle model; B, round handle model. (*b*) Diagram showing position of tungsten shielding and method of source mountings. (Reproduced from Delclos *et al* (1978) by permission of the American Cancer Society Inc.)

outlining milligram–hours and application times used for the Fletcher system are used with other applicators since applicator design will give different doses to the bladder, rectum and vaginal mucosa for the same nominal dose to Point A. This is particularly important for any applicator system which does not use shielding in the colpostats or for applicators in which the source axis is along the axis of the vagina, rather than at right angles to the intra-uterine tube, as is the case with the Henschke applicator.

Another simple manual after-loading technique using J-type caesium tubes (Amersham International plc) has been described by Haybittle and Mitchell (1975). This applicator consists of clear PVC uterine tubes, closed at one end with a shaped bung whilst the other end is fitted with a stainless steel threaded end to take a cap which maintains the source in position. The three sizes of ovoids are made of rubber fitted onto a tube which again is fitted with a stainless steel thread which will take an end cap. The caesium tubes themselves are loaded in clear PVC tube sealed at one end with PVC rod and having the stems fitted with PVC rod, followed by a coded bung which is sealed in position before insertion. In practice, the ovoids are placed in position in theatre and the sources are loaded into the tubes at a later time and secured in position by the end caps.

The main advances in after-loading techniques followed the introduction of miniature caesium sources in the 1960s. Horwitz *et al* (1964) described an after-loading system using 10 mm sources loaded with microspheres of caesium-137. These sources enable flexible source trains to be constructed which can then be inserted into the applicators. Redpath *et al* (1976) described the adaptation of the Edinburgh application using a single line source in an after-loading technique. Following this trend, a manual after-loading system has been developed commercially by Amersham International plc (Bateman *et al* 1983, Mould and Hobday 1984). The applicator dimensions of this system follow closely the recommendation of the Manchester system. Figure 7.10 shows the Amersham system of plastic applicators, assembled using 'spacers' and 'washers'. The uterine and ovoid tubes are made of flexible PVC tubing and the ovoids, spacers and washers of high-impact polystyrene. The applicators, all of which are impregnated with barium sulphate to enable visualisation on radiographs (Ellis and Taylor 1982), are made in separate sections which can be locked together.

The source trains used in these applicators consisted of seed sources of caesium-137 incorporated into three small glass beads, interspersed

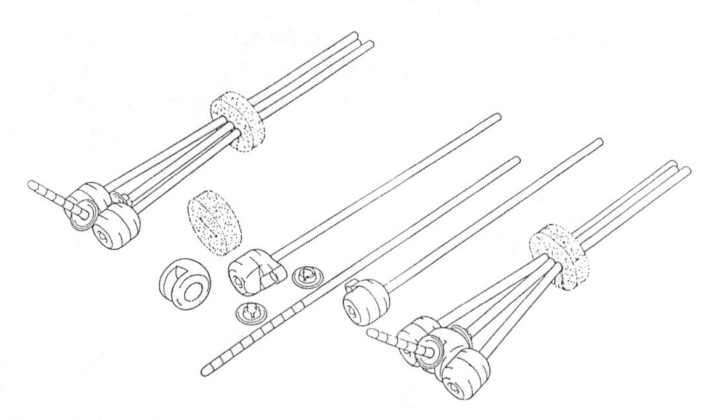

Figure 7.10 Components of the Amersham disposable applicator after-loading system. (Reproduced by kind permission of Amersham International plc.)

with inactive spacers. The spacers and active beads are held in position in a closed end of a flexible source holder by a flexible spring (figure 7.11). The other end of the spring is a plug which acts to close the holder and also identifies the source train. The source strength of each bead in the train has been chosen so that the various available source trains (figure 7.12) simulate the typical loadings used in the Manchester dosage system. To achieve the desired distribution it is arranged that the lower end of the uterine source train coincides with the external os and this is achieved by ensuring that the front locking flange is at the correct position on the uterine central tube. As with other manual after-loading techniques, the applicators are inserted into the patient in theatre and source trains then inserted once the patient has returned to the ward. It is envisaged that these applicators will be used on a disposable basis.

To simplify the use of the system a set of dose tables have been derived by Mould and Hobday (1984) for a 'standard insertion' for the various combinations of vaginal and uterine applicators—a standard insertion being defined as one in which the angle, α, between the uterine and vaginal axes is 120°, and the angle, β, between the uterine and vaginal planes is 0°. Guide-lines have also been given to indicate the range of α and β and the maximum amount of ovoid slippage below the 'locking washer' at the external os, for which the table applies for different degrees of dose accuracy at Point A. For the standard insertion the relative dose rate contribution to Point A for each type of uterine or tandem source train or two ovoids is as shown in table 7.8, column five.

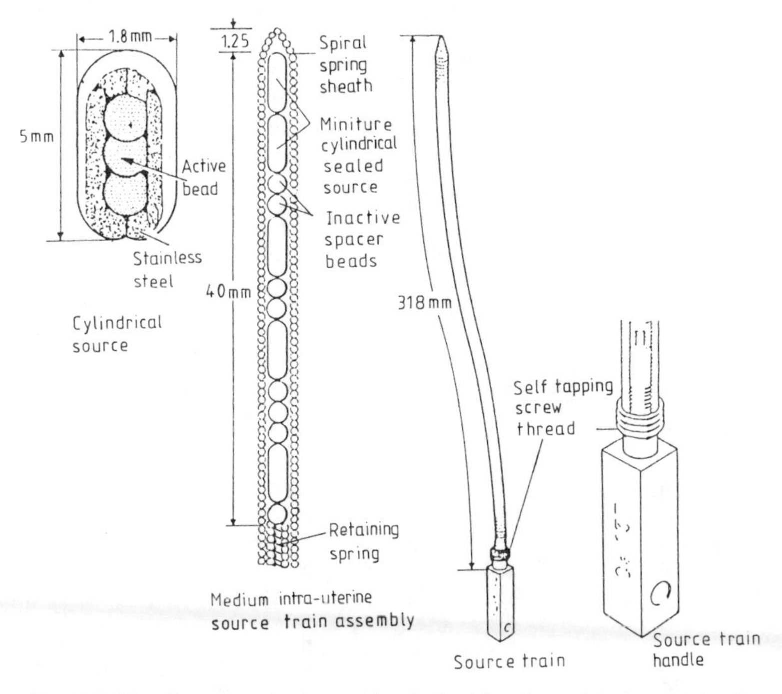

Figure 7.11 Source train assembly design for Amersham manual after-loading system. (Reproduced by kind permission of Amersham International plc.)

These derived tables only apply for the nominal content activities of the sources used when making the calculations. When sources are supplied by the manufacturer correction has therefore to be made to these tables to account for the actual activity within the source which will be different from the nominal content activity. Sources supplied with a certificate stating the exposure rate at 1 m from the source are corrected by deriving the factor which is the ratio of the actual exposure rate to the value of the exposure rate used for the calculation given in figure 7.12. If the source is specified in terms of equivalent activity, then the exposure rate must be determined by using the exposure rate factor recommended by the manufacturer. To comply with the ICRU recommendations the source strength is also specified in terms of air kerma rate to air.

Suppose the source trains arrive with the exposure rates at 1 m as shown in column three of table 7.8, then the correction factor in column

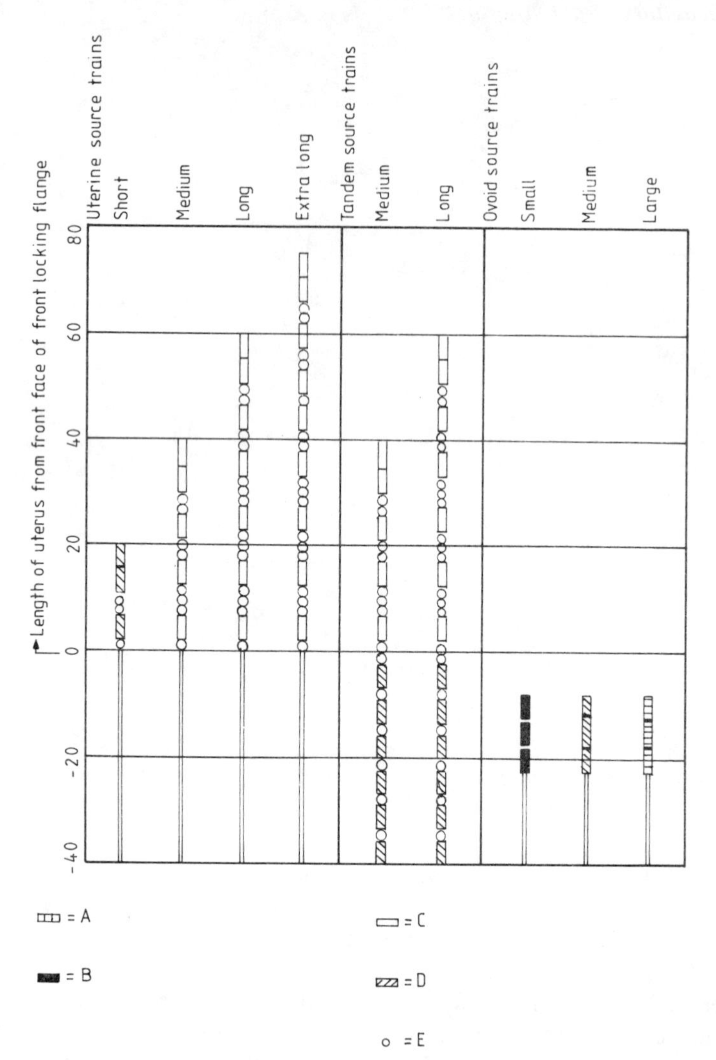

Figure 7.12 Source train loading pattern specifications for Amersham manual after-loading system. Description of individual caesium-137 sources. A: nominal content activity = 17.0 mCi (627 MBq); exposure rate at 1 m = 5.52 mR h^{-1}; air kerma rate at 1 m = 48.2 μGy h^{-1}. B: nominal content activity = 13.2 mCi (488 MBq); exposure rate at 1 m = 4.29 mR h^{-1}; air kerma rate at 1 m = 37.5 μGy h^{-1}. C: nominal content activity = 11.3 mCi (418 MBq); exposure rate at 1 m = 3.68 mR h^{-1}; air kerma rate at 1 m = 32.1 μGy h^{-1}. D: nominal content activity = 15.1 mCi (558 MBq); exposure rate at 1 m = 4.90 mR h^{-1}; air kerma rate at 1 m = 42.8 mGy h^{-1}. E: inactive spacer bead. (Reproduced by kind permission of Amersham International plc.)

Table 7.8 Dose rates at Point A due to nominal and actual source trains.

Source train	Exposure rate per seed Nominal	Exposure rate per seed Actual	Correction factor	Dose rate in tissue to Point A ($Gy\,h^{-1}$) Nominal	Dose rate in tissue to Point A ($Gy\,h^{-1}$) Actual
Uterine tubes					
short	4.90	5.72	1.167	0.285	0.333
medium	3.68	4.20	1.141	0.320	0.365
long	3.68	4.18	1.136	0.340	0.386
extra long	3.68	4.18	1.136	0.370	0.420
Tandems					
medium	3.68 4.90	4.15 5.56	1.128	0.500	0.564
long	3.68 4.90	4.15 5.56	1.128	0.520	0.587
Ovoid assembly					
small	4.29	4.89	1.140	0.180	0.205
medium	4.90	5.57	1.137	0.200	0.227
long	5.52	6.48	1.174	0.220	0.258

Table 7.9 Dose rates to Point A for combinations of uterine tubes and ovoid assemblies.

Uterine tube	Ovoid assembly	Dose rate in tissue to Point A ($Gy\,h^{-1}$)
Short	Small	0.538
	Medium	0.560
	Large	0.591
Medium	Small	0.570
	Medium	0.592
	Large	0.623
Long	Small	0.591
	Medium	0.613
	Large	0.644
Extra long	Small	0.625
	Medium	0.647
	Large	0.678
Tandem medium		0.564
Tandem long		0.587

four enables the nominal dose tables to be recast to give the actual dose rate in tissue to Point A from the various constituent elements of the Amersham after-loading system.

To determine the dose rate in tissue at Point A for the various combinations of uterine tube and ovoid assembly the contribution of each component is summed as detailed in table 7.9.

It must be remembered that such a table is only applicable to the date of measurement indicated on the test certificate. For clinical purposes it is useful to recast table 7.9 so that it gives the time of treatment in any calendar year to give a prescribed dose to Point A for any particular combination of uterine tube and ovoid assembly.

7.4.3 Remote-controlled after-loading (low dose rate)

To reduce exposure to all staff after-loading systems have been developed which enable sources to be transferred remotely into the applicators for treatment from a lead-shielded storage safe. These sources are then returned automatically to the safe at the end of treatment and at any time when the patient requires nursing care. These techniques imply that the patient will have applicators inserted as in the manual after-loading systems and then once the patient has returned to the ward the applicators are connected to the treatment machines. It is essential that in this type of system safety features are incorporated into the design to ensure that in the event of mains failure the sources are automatically retracted to a safe condition. Three main systems have been developed to achieve low-dose-rate after-loading, each using a different method of source transfer and applicator system.

Preformed source trains, which are transferred from the sealed-source safe into the applicator using drive cables, provide the basis of one after-loading system—the CGR 'Curietron'. In this unit the source train consists of caesium-137 sealed sources and spacers loaded into a stainless steel spring and closed at one end. The ends distal to the sources are sealed with a hook which serves as an identification of the source train and provides the point of attachment to the drive cable. A maximum of four source trains may be loaded into a Curietron at any one time, each of which is connected to an independently controlled drive cable. The movements of the source train are remotely controlled from the central console which is sited outside the treatment room. Each separate channel has an independent timer which enables that particular source train to be retracted after the appropriate time for that channel.

Due to the use of source trains which are sealed and loaded with predetermined activity distribution at the time of manufacture, a stock of source trains is necessary and it is therefore important that the loading patterns are selected carefully prior to their manufacture to suit local requirements. The source trains are typically stored in the main sealed-source safe and a portable safe is used to transfer the sources required for a particular treatment between the main safe and the Curietron in the ward.

Various types of applicators have been developed by the manufacturers for use with this equipment. With the Curietron, automatic retraction of sources is achieved under battery power in the event of a power failure. The use of such a system has been described by numerous authors including Hope-Stone *et al* (1981) and Jackson and Davies (1983).

Another system which uses sources driven by cable is that produced by Buchler GDBH. Rather than use a source train, the manufacturers have chosen to make the source oscillate within the uterine tube. This can be used by itself to produce the required distribution in the single-channel version or in conjunction with two fixed sources in the three-channel version. When a single source is used the desired dose distribution is achieved by using a program disc whose size and shape determines how long the source is maintained in the various positions of the central tube. During treatment, the source which is mechanically driven and controlled by the program disc moves with constant frequency within the applicator over a path length of up to 200 mm. In the three-channel system, designed to produce typical Manchester or Houston distributions, two fixed ovoid sources remain in position for the treatment. The main appeal of this system is that the many different source trains of the Curietron are not required, since the various distributional patterns required can be obtained by using the suitable program disc.

The third low-dose-rate after-loading system, the Selectron developed by Nucleton, uses source trains made up of active and inactive spherical pellets. The unit contains its own sealed-source safe and in the setting-up procedure the sources are sorted magnetically and arranged into trains of 48 pellets per channel by the machine under microprocessor control to give the required dose distribution. The source trains are then loaded into an intermediate safe and, once loaded and the treatment time entered for the appropriate channel, the key is removed from the programming console in the top of the unit, thereby making it impossible to tamper with this source loading during the course of the treatment. In clinical practice the applicators, which are inserted into the patient in theatre, are connected to the unit when the patient returns to the ward. Once connected, the source trains are loaded into and withdrawn from the applicators by compressed air. During treatment the machine is controlled from a panel outside the room, which shows the treatment time remaining, the channels being used in treatment and any error conditions. In the event of any fault developing in the mains or compressed air supply, the sources are automatically returned to the intermediate safe, using the back-up batteries or the reserve air tank. In normal use, the unit withdraws the sources prior to staff entering the treatment room, thereby enabling nursing procedures to be carried out without exposing staff to radiation. Since the Selectron is a six-channel unit, it is possible to treat two patients simultaneously, each with three applicators, and the system is so designed that each patient can be treated independently of the other. The range of applicators available with this system enables the Houston, Stockholm and Paris systems to be reproduced, together with specially designed applicators. The ability to arrange the active sources in any of 48 possible positions enables a degree of flexibility in the treatment of cancer of the uterine cervix and body of uterus, as well as lesions in the vagina.

The clinical use of low-dose-rate after-loading systems has been described by many authors (Hope-Stone *et al* 1981, Wilkinson *et al* 1980, 1983, Corner *et al* 1982). Depending on the activity of the sources used, which typically are 20, 30 or 40 mCi caesium-137 pellets, the overall treatment times have varied between 15 and 50 hours. Some workers have deliberately worked at the dose rates advocated by conventional techniques, whilst others have considered that from the point of view of patient comfort the maximum treatment time should be reduced to a minimum of 30 hours. Using these reduced treatment times, allowance has to be made for the biological effect experienced when using dose rates of the order of 150–170 $\mathrm{cGy\,h^{-1}}$. Initially, the overall dose has been reduced by the order of 10% from those used for 50-hour treatments (Wilkinson *et al* 1980), although the use of TDF or CRE data would indicate a reduction of the order of 30%. However, work is still progressing to establish the optimum dose to achieve the same cure rate as was obtained by the traditional techniques.

7.4.4 Remote after-loading systems (high dose rate)

High-dose-rate systems use sources of much greater activity so that the treatment times are reduced from hours to minutes. The use of these activities does mean, however, that suitable protection has to be arranged for staff and the source safe itself needs to afford a greater degree of protection and thus is consequently much heavier than with low-dose-rate systems. Since these high activities yield increased dose rates, the management of the patient treatment is different. Rather than the treatment being carried out in ward areas, it is customary with high-dose-rate after-loading systems to have a purpose-built treatment suite or to suitably adapt an existing teletherapy room, which of necessity has adequate protection. In the treatment suite it is desirable to have a patient preparation area where the applicators can be inserted into the patient, a treatment room where the irradiation is given and a recovery room all adjacent to each other. Due to differences in the dose rate, the treatment is fractionated over a number of days which necessitates the patient undergoing up to nine theatre visits instead of the two or three with low-dose-rate systems. The increased risk associated with multiple anaesthetics is seen by many to be a disadvantage of the system. Many workers, however, consider that the benefits to be gained by having greater precision of treatment due to the sources remaining fixed during the insertion times, together with the fact that patients can be treated as outpatients, outweigh the disadvantages.

The system which has been used in the UK for more than 20 years is the Cathetron, although in recent years both the Buchler system and the Selectron have been adapted for use with high-activity sources. First described by O'Connell *et al* (1965, 1967), the Cathetron utilises high-activity cobalt-60 source trains. The applicators, in the form of stainless steel catheters, are inserted into the patient in the theatre and are clamped in position, the clamps being attached to a longitudinal bar fixed to the operating table. Once in the treatment room, the catheters are connected by flexible nylon-lined supply tubes to the source storage container, through which the source pencils are driven into the catheters. Depending on the installation, the flexible drive cables pass under the floor from the treatment room to the control unit, which has to be situated outside the treatment room. During the five- to fifteen-minute treatment, the patient is viewed by close-circuit television or direct observation through a treatment window and clinical parameters such as pulse and respiration are monitored. The sources are driven out independently by three separate electronic motors and are automatically returned to the storage container at the end of the preset time.

Whilst the system allows the source position to be monitored at any instant of time, various authors (Liversage *et al* 1967, Twiss and Bradshaw 1970) have reported that 'snaking' of the drive cable within the supply tubes connecting the control to the storage container can result in the apparent source position, as indicated on the control panel, being in error by 3 mm. This effect can lead to significant error in dose, although in recent years the manufacturers have introduced spring loaded source pencils in an attempt to rectify this problem.

Santhamma and Das (1978) carried out experiments using TLD to determine the optimum source arrangement to achieve a 'Manchester distribution', although it should be noted that the vaginal colpostats are not in the true Manchester orientation with respect to the central uterine tube. These results were used to verify theoretical calculations which took account of the filtration at the ends of the sources, the effective filtration of the cobalt-60 gamma-rays in the sheathing of the source capsule, the thickness of the helical steel spring surrounding the source element and the wall of the stainless steel catheter. Earlier, Liversage *et al* (1967) devised source loadings to achieve the same relative contributions to Point A from the central tube and ovoids as in the Manchester system. With the activity used, the standard treatment of 1000 cGy to Point A is given in five to six minutes. In producing isodose curves allowance was made for the effects of oblique screenage. Since the Cathetron source pencil comprises a number of short capsules, the gamma-ray absorption through the hemispherical ends increases with obliquity rather less than would be the case if the sources were contained in a long steel cylindrical tube.

To obtain an estimate of the effective wall thickness, suppose the rate at 20 cm from a cobalt-60 point source is X_n. The source is now placed inside a long cylindrical absorber of wall thickness d and of linear attenuation coefficient μ, and an exposure rate X_θ measured 20 cm from the source at an angle θ to the cylinder.

Since

$$X_\theta = X_{\mathrm{n}} \exp(-\mu d \operatorname{cosec} \theta)$$

then, by measuring the exposure rate at $\theta = 30°$ and $\theta = 90°$, X_{30} and X_{90}, and taking the ratio, μd can be determined since $X_{30}/X_{90} = \exp(-\mu d)$. Using this approach, Liversage derived the effective value of μd, from which it was possible to calculate the effective activity of the source. Verification of these calculations was carried out using a cadmium sulphide crystal attached to a dose plotter. In determining the dose given to the patient it is necessary to consider the dose contributions received whilst the source pencil is in transit from the storage container to the treatment position at the end of the catheter. An estimate of this dose can be achieved by using the formula given by Walstram (1965):

$$N_t = (A\Gamma_\delta / rV) \int_{\phi_1}^{\phi_2} \exp\left(-\sum \mu_i d_i / \cos\phi\right) \mathrm{d}\phi$$

where Γ_δ is the specific gamma-ray constant, μ_i is the linear attenuation coefficient of the absorber of thickness d_i and ϕ_1 and ϕ_2 are the angles subtended at the beginning and the end of treatment. Using this approach, and verifying by experiment, Liversage determined the dose due to a 5 Ci source being transferred in and out 15 times to be about 130 R for a 4.5 s transfer time.

7.5 Rectal and Bladder Dose Measurement

In treating gynaecological malignancy a limiting factor is the radiation dose which can be tolerated by the bladder and rectum. It is therefore of utmost importance for the radiotherapist to know the doses at these sites when the sources are *in situ*. These can be measured in practice using a suitable detector at the time of operation. This knowledge enables the radiotherapist to decide if the insertion is satisfactory or whether or not the sources need to be replaced or the applicators repositioned in order to bring the rectal dose down to between 60% and 70% of the dose to Point A.

Initially, with manual loading intracavitary techniques Fletcher *et al* (1952, 1953) described the use of a scintillation counter and a small ionisation chamber connected to a direct-reading meter to determine the dose along the base of the bladder and the anterior rectal wall. In estimating dose, the probe has to be positioned carefully and without pressure if a true reading of the rectal and bladder dose is to be achieved. It must also be remembered that changes in patient position will affect the dose. In many centres the applicators are introduced with the patient in the lithotomy position. Joelsson and Bäckström (1969) have reported that the bladder dose can increase by 12% when the patient is supine and that changes in the dose rate on the rectal wall can be as much as +26% over the course of a 24-hour insertion. These changes are attributed in part to positional changes in the source position and also to changes in the gauze packing during the course of treatment.

Various detectors have been used over the years to measure intracavitary doses. With manual techniques, detectors capable of quickly and accurately measuring dose rates of the order of 0.5 $\mathrm{cGy\,s^{-1}}$ (30 $\mathrm{cGy\,h^{-1}}$) are required and thus it is possible to use relatively small ionisation chambers. However, with after-loading systems, where the philosophy is to keep exposure dose to staff to a minimum, the sources used for treatment are not inserted into the patient at operation; thus in order to determine rectal dose it is necessary to insert lower-activity sources into the applicators in theatre, thereby mimicking the intended sources for treatment.

The use of low-activity sources means that it is necessary to measure dose rates of the order of 0.05 $cGy\,s^{-1}$ (3 $cGy\,h^{-1}$) and this places severe limitations on the measuring system. In order to achieve a usable response time the size of the ionisation chamber needs to be of the order of 1 cm^3, which is large for clinical use. Other detectors, such as solid state and photovoltaic detectors, have therefore been evaluated. These detectors are often temperature- and energy-dependent, which means that they have to be used with great care. Sheldon *et al* (1969) described a technique in which the dose was assessed in theatre using a rigid scintillation probe to obtain doses at 1 cm intervals. The true rectal dose, however, is more accurately determined by inserting a rectal catheter loaded with TLD and leaving it in position throughout the course of treatment.

In an attempt to assess the optimum method of determining rectal dose, Corner *et al* (1982) carried out a series of measurements on 40 patients who were treated by a low-dose-rate after-loading Selectron system. Their technique was to use low-activity F-type tubes (from Amersham International plc), a single source being used to simulate an ovoid and one or two sources spaced in such a way so as to simulate source trains. A semiconductor detector was used for this work, although the authors acknowledge that this detector showed instability at low dose rate and had a temperature coefficient of 0.05 $cGy\,(^{\circ}C)^{-1}$ (this effect was partly overcome by using a Perspex sheath over the probe). In order to estimate the dose rate used in treatment a factor was used to correct for the difference in treatment and test loadings, the factor being derived from measurements. The uncertainty in the estimated dose rate was $\pm 20\%$. Using this technique, the dose to the rectal wall was determined and then compared to calculated doses derived from localisation films taken with rectal markers in position and also from direct measurement made on patients during treatment. This series showed that the rectal dose rates estimated in theatre using test sources were 6% higher than those measured in treatment and those calculated from radiography were 4% lower than the treatment rectal dose rates.

7.6 Combined Intracavitary and External Beam Therapy

The rapid fall-off in dose around the intracavitary applicators prevents the parametrium and the regional lymph nodes from receiving adequate treatment, even though the colpostats may be widely separated. Therefore, it is often the practice to supplement intracavitary treatment with external radiation. The relative times at which the internal and external therapy are given depend on the philosophy of treatment regime adopted, which is a function of the staging of the disease. For certain clinical conditions the intracavitary therapy is carried out as the last part of the treatment regime, by which time it is hoped that the macroscopic disease in the pelvis is abolished and the intracavitary component of the treatment is used to produce final obliteration of the disease. In prescribing dose for this technique, the dose given by the intracavitary therapy must be influenced by the tolerance of the bladder and rectum, as well as the requirement to give the tumour as high a dose as possible. It must also be recognised that in this technique the dose to the lining of the vaginal mucosa will be lower.

The other approach to combined intracavitary and external beam therapy is to perform the intracavitary procedures at the beginning of the treatment and then to give external beam therapy to irradiate the parametrium and regional lymph nodes, but using a metal block to shield the volume irradiated by the intracavitary therapy. With megavoltage irradiation this is often achieved by the use of a two-field antero-posterior technique with a wedge filter placed centrally, the wedge being designed to reduce the contribution of x-rays at points near to the sealed sources, whilst at the same time raising the external beam dose to points in the parametrium. In designing such a wedge, consideration has to be given to the rapid fall-off in dose of the intracavitary therapy, the maximum dose to Point A and the desired dose to the rest of the pelvis (figure 7.13). The wedge itself has to be capable of being positioned so that it can be aligned with the thickest portion over the position occupied by the uterine sources. This can be achieved by the use of a template (figure 7.14).

The use of such wedges on linear accelerators has been described by many authors (Tranter 1959, Blomfield 1961, Cowell and Laurie 1967, Jameson and Trevelyan 1969, Jones *et al* 1972, Glazebrook 1974). To accommodate the different treatment requirements for various clinical conditions, wedges have been designed which allow different amounts of radiation to be transmitted through the wedge. If a cobalt-60 teletherapy unit is used for external treatment, it is possible to use a simple straight-sided block rather than a shaped block, since the source size of such units is greater than for a linear accelerator and they thus create their own penumbra.

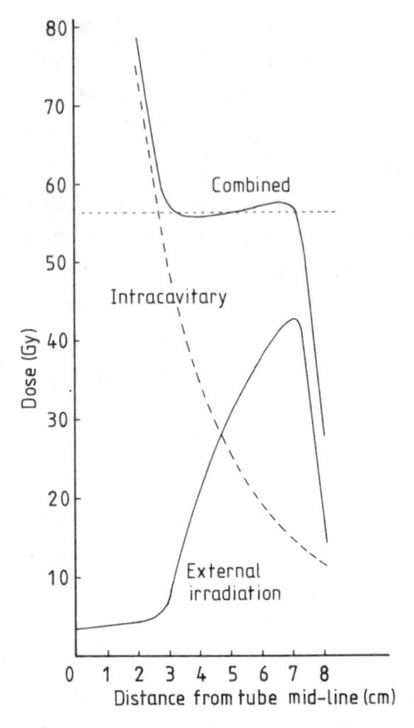

Figure 7.13 Combination of dose from intracavitary and external beam therapy. (After Tranter (1959) by kind permission of *British Journal of Radiology*.)

Isodose distributions showing the combined intracavitary and external beam therapy have been given by Blomfield (1961), Cowell and Laurie (1967) and Jones *et al* (1972). These distributions show the dose within the plane through the level of Point A. However, it is necessary to know the dose to the regional lymph nodes and this is defined by the placing of the block within the coronal plane of the patient. Various techniques have been adopted to define the exact position of the intracavitary sources, by using 'inactive' gold grains, for example, or mechanical aids (Delclos 1984). Techniques vary as to the extent of shielding required, some radiotherapists preferring to shield the whole of the central position of the opposed field, whilst others limit the shielding to the actual volume irradiated by the sealed sources (Cowell and Laurie 1967, Fletcher 1980), the upper edge of the block being positioned so that it shields a volume up to 5 cm above the line which is a tangent to the femoral heads. In this technique the external beam is rotated 45°, so that the diagonal of the beam follows a line 1 cm above the tangent to the femoral heads. In order to account for the likelihood of the two radium insertions not being identical and the uterus not being in the mid-line of the patient, it is necessary to ensure that the block can be adjusted to compensate.

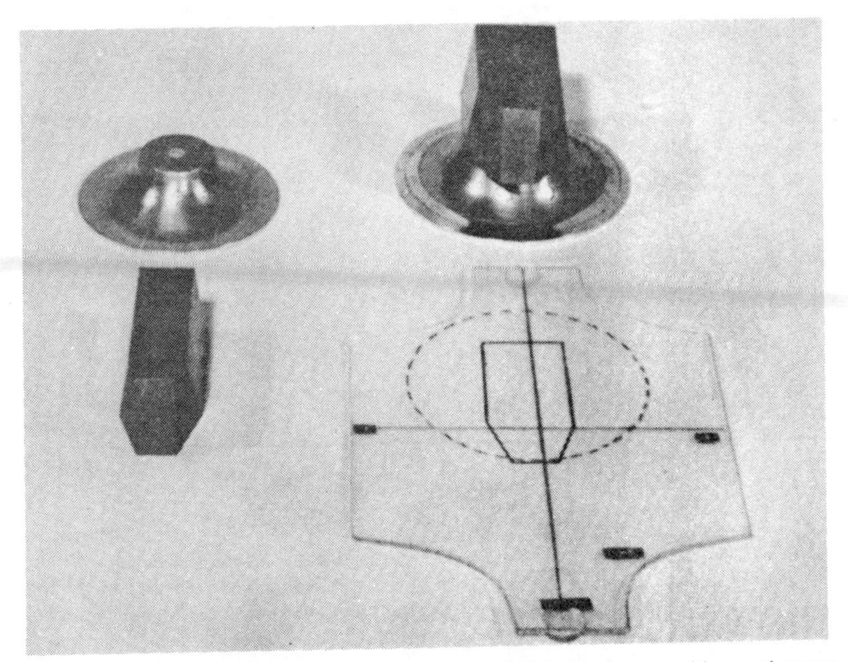

Figure 7.14 Wedges for use in combined intracavitary and external beam therapy. (From Jones *et al* (1972) by permission of *British Journal of Radiology*.)

Positioning the wedge is extremely important, since any error can result in marked differences in dose across the pelvis. Johnsson and Nordberg (1975) demonstrated the difficulties met in obtaining an adequately summed dose distribution in the treatment volume when external therapy is given after intracavitary treatment. For the ideal distribution of dose the position of the uterus must be the same with and without the applicator inserted. However, in many cases, they found that the position of the uterus was not the same and that since the block position is determined from the radiographs of the insertion, the volume shielded is not necessarily the volume irradiated by the sources. The effect of such a change in uterine position results in areas of over- and underdosage (figure 7.15).

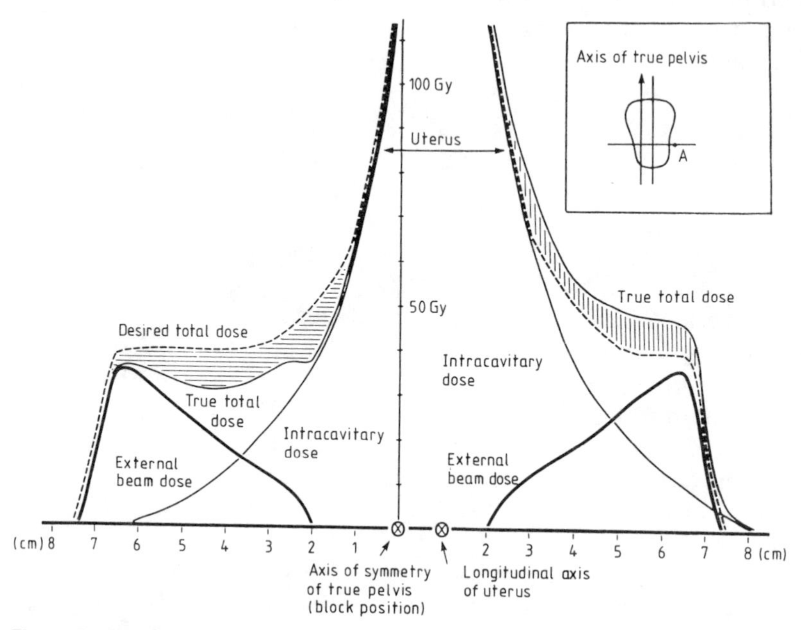

Figure 7.15 Distribution of dose in the uterus and surrounding tissues with the applicator placed 1 cm laterally to the left and the shielding block placed exactly corresponding to the axis of symmetry of the true pelvis. (From Johnsson and Nordberg (1975) by permission of *Acta Radiologica*.)

In addition, it is necessary to consider the summed dose distributions from all the intracavitary treatments. Whilst the doses at points fixed to the uterus can be adequately screened, Point B and the edge of the true pelvis are likely to receive unequal doses and this must be taken into account in choosing the wedge, together with inaccuracies which can occur in the day-to-day setting up of external beam therapy.

However, a more fundamental inaccuracy occurs from the use of radiographic techniques to derive the true position of the sources within the pelvis. The limitation of these techniques, discussed below in §8.3, means that an accuracy of the order of ± 3 mm can be achieved at best. In addition, the positioning of a block on the lead tray cannot be achieved to better than ± 1 mm which, when magnified to the mid-line of the patient, can result in a further positional uncertainty of ± 2 mm. Such inaccuracies can lead to significant regions of overdosage and underdosage as shown. It is, therefore, perhaps open to question if the effort involved in devising such complicated block construction is valid.

7.7 Absorbed Dose and Biologically Effective Distributions

Typically, dose distributions obtained by calculating the dose around sources are usually shown in the plane defined by the axis of the uterine tube (figure 7.16). However, it must be remembered that the distribution is a pear-shaped three-dimensional volume (figure 7.17) which is often at an inclined angle within the pelvis. When adding external beam therapy to intracavitary therapy, the position of the shielding block must be matched to the intracavitary therapy, which for an opposed pair of external beams will not necessarily be the shape of the distribution shown in the plane of the uterus but rather its projection onto a horizontal plane through the patient and parallel to the radiograph. It is also customary to add the dose given by the external beam to that due to the insertions at different sites within the pelvis. However, it must be remembered that since the dose rate varies markedly throughout the volume, the biological effect due to the irradiation will be different in different sites due to the variation in biological effect with dose rate. This is maximal in the range 0.02 to 1.6 $\mathrm{cGy\,s^{-1}}$, as shown in §1.3.1. Thus this effect is important when considering doses throughout the pelvis where the dose rate falls rapidly from the centre of the pelvis to the pelvic side walls.

We have considered above the need to reduce the dose prescribed to Point A if the dose rate is increased from 50 to 150 $\mathrm{cGy\,h^{-1}}$. Since the dose rate at Point B is about 1/3 that of Point A, the biological effect at that point may change differently to that at Point A when the dose rate is changed from 50 to 150 $\mathrm{cGy\,h^{-1}}$. The case can therefore be made to produce a dose distribution in terms of isoeffect rather than isodose.

From § 1.3.2 we known that isoeffects can be derived by using the TDF or CRE relationships or the α/β ratio of the linear quadratic model. By considering the TDF equation (equation (1.21)) we can see that for a given insertion time the TDF for different positions in the pelvis, TDF_x

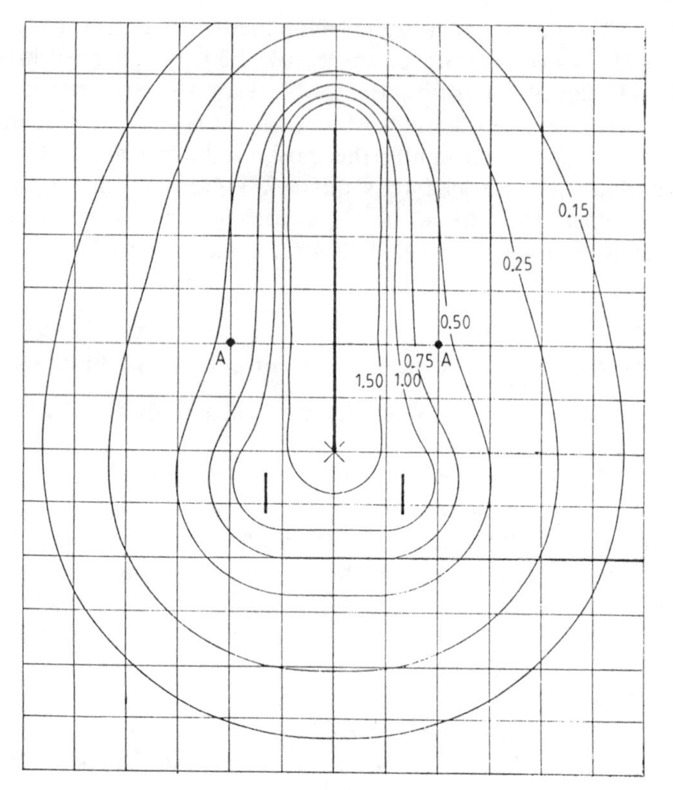

Figure 7.16 Distribution in plane of uterus for a long intra-uterine applicator and two medium ovoids (all dose rates relative to the dose rate at Point A, given in Gy h^{-1}). Scale: 1 square = 1 cm. (Reproduced by kind permission of Amersham International plc.)

due to the different dose rates, is related to the TDF_A at Point A by the relationship

$$TDF_x = TDF_A(r_x/r_A)^{1.35}$$

where r_x is the rate at point x and r_A the rate at Point A.

By normalising the dose distribution to the value at Point A we can see that the isoeffect distribution is different from the isodose curve (figure 7.18). A similar approach can be used to combine intracavitary and external beam therapy to define the overall treatment. This approach is particularly important when combining external radiotherapy with high-dose-rate brachytherapy, as has been shown by Joslin *et al* (1972).

Many different treatment regimes are used in the treatment of carcinoma of the cervix. In order to compare their biological effectiveness use can again be made of TDF or CRE systems for the total treatment, taking into account the gaps between the brachytherapy and external beam therapy using the recommended factor of Orton (1974) or Kirk *et al* (1975a, 1977a). Using the TDF system, the overall effect of the treatment is given by

$$\mathrm{TDF}_{(\mathrm{tot})} = \mathrm{TDF}_{(\mathrm{phase}\ 1)} + \mathrm{TDF}_{(\mathrm{phase}\ 2)} + \mathrm{TDF}_{(\mathrm{phase}\ 3)} + \cdots .$$

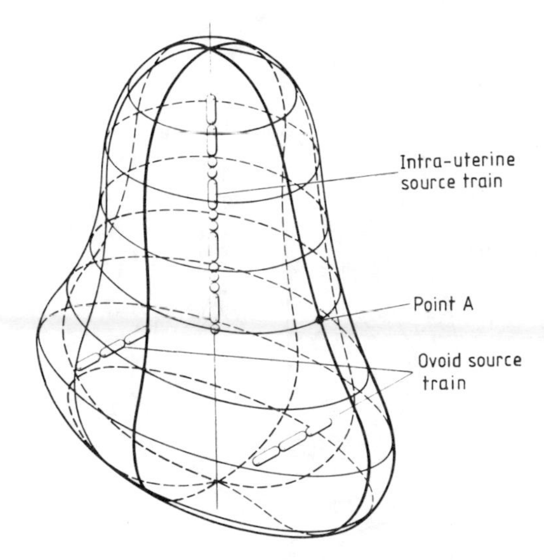

Figure 7.17 Isodose surface through Point A for Amersham manual after-loading system. (Reproduced by kind permission of Amersham International plc.)

In summing the TDFs a correction, the decay factor d, has to be made to account for the gap between treatments. The decay factor proposed by Orton and Ellis (1973) is of the form

$$d = [T/(T+R)]^n$$

where T is the total time elapsed in days between the first day of treatment and the rest period, R is the duration of the rest period in days and n is a constant. The constant, n, was originally set at 0.11 but has subsequently been corrected to 0.169 to preserve mathematical consistency (Goitein (1976), Kirk *et al* (1975a) and acknowledged by Ellis (1985)).

Using the TDF relationships and the above gap correction factor, it is possible to compare the biological effects of different treatment regimes, as shown in the following example. In this example we make the assumption that the pelvic nodes receive 15% of the dose to Point A for each insertion.

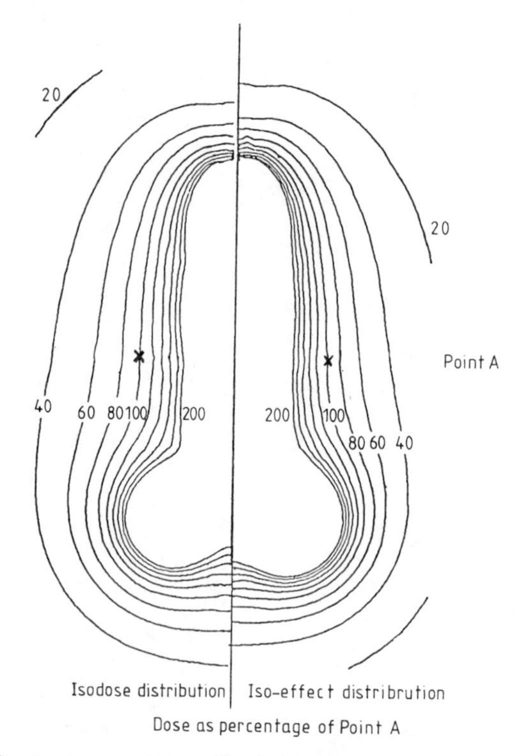

Figure 7.18 Isodose and iso-effect distributions around a typical gynaecological insertion. Left-hand side is the isodose distribution, right-hand side the iso-effect distribution.

Example 7.4

Compare the TDF for a treatment regime of two insertions, each giving 30 Gy in 36 hours to Point A, separated by one week, followed by external beam therapy one week later giving 15 Gy to Point A and 40 Gy to the pelvic nodes in 20 fractions in a total of four weeks (26 days), with a regime using a low-dose-rate after-loading system of two insertions of 25 Gy to Point A at a dose rate of 175 $cGy\,h^{-1}$, followed one week later by the same external beam therapy.

(i) *For lower-dose-rate intracavitary regime.* Dose rate to Point A = 53.6 cGy h^{-1}, dose rate to pelvic nodes = 8.0 cGy h^{-1}. Using equation (1.21), TDF for each insertion to Point A = 57.6, pelvic nodes = 4.4. Decay factor for rest period between first and second insertion = $(2.3/7)^{0.169} = 0.829$. Decay factor for rest period between second insertion and start of external therapy = $(9.3/14)^{0.169} = 0.933$.

The TDF for external therapy to Point A (20 fractions of 75 cGy in 26 days) = 14.6. TDF for external therapy to pelvic nodes (20 fractions of 200 cGy in 26 days) = 66.2. The overall TDF at Point A = $[(57.6 \times 0.829) + 57.6] \times 0.933 + 14.6 = 112.9$. The overall TDF at pelvic nodes = $[(4.4 \times 0.829) + 4.4] \times 0.933 + 66.2 = 73.7$.

Table 7.10 Comparison of the isoeffect of four regimes for treatment of carcinoma of the cervix using TDF relationships.

Regime	Technique	TDF at Point A	TDF at pelvic nodes
A	Two insertions each of 30 Gy in 56 hours to Point A, separated by one week, followed one week later by external therapy of 15 Gy to Point A and 40 Gy to pelvic nodes in 20 fractions in 26 days (4 weeks).	112.9	73.7
B	Two insertions each of 25 Gy in 14 hours to Point A, separated by one week, followed one week later by external therapy of 15 Gy to Point A and 40 Gy to pelvic nodes in 20 fractions in 26 days (4 weeks).	123.3	74.6
C	Two insertions each of 20 Gy in 24 hours to Point A, separated by one week plus a third insertion of 20 Gy in 24 hours two weeks later, followed one week later by external therapy of 15 Gy to Point A and 40 Gy to pelvic nodes in 20 fractions in 26 days (4 weeks).	120.2	74.5
D	External beam therapy of 46 Gy in 23 fractions in 31 days to Point A (pelvic nodes receive 95% of this dose), followed by a single insertion of 30 Gy in 36 hours after a 4-week gap.	134.7	67.7

(ii) *For higher-dose-rate intracavitary regime.* Dose rate to Point A = 175.0 cGy h^{-1}, dose rate to pelvic nodes = 26.3 cG h^{-1}. Using equation (1.21), TDF for each insertion to Point A = 72.6, pelvic nodes = 5.6. Decay factor for rest period between first and second insertion = $(0.6/7)^{0.169} = 0.660$. Decay factor for rest period between second insertion and start of external therapy = $(7.6/14)^{0.169} = 0.902$. The TDF for external therapy to Point A (20 fractions of 75 cGy in 26 days) = 14.6. TDF for external therapy to pelvic nodes (20 fractions of 200 cGy in 26 days) = 66.2. The overall TDF at Point A = $[(72.6 \times 0.660) + 72.6] \times 0.902 + 14.6 = 123.3$. The overall TDF at pelvic nodes = $[(5.6 \times 0.660) + 5.6] \times 0.902 + 66.2 = 74.6$.

Such a technique provides a method of assessing the biological effect of different treatment regimes. A comparison of four commonly used regimes is given in table 7.10.

Further methods of estimating the biological effect of combined intracavitary low-dose-rate irradiation with external beam therapy have been given by Ellis and Sorensen (1974) using the NSD approach, Kirk *et al* (1975b, 1977a) using the CRE relationships and Dale (1985) using linear quadratic relationships.

7.8 Neutron Intracavitary Therapy

One of the early reports on the use of californium-252 for gynaecological intracavitary brachytherapy was given by Castro *et al* (1973). In their work californium-252 tubes were made to fit the standard after-loading Fletcher/Suit colpostats and the applicators were loaded on the basis of replacing 1 mg of radium-226 by 0.45 μg of californium-252, thus the uterine tube contained (7.5 + 7.5 + 5.0) μg of californium-252 and the vaginal colpostats 10 μg each. The use of such relatively large quantities of neutron-emitting material raised major problems in terms of staff safety and these have meant that few centres have continued with this work.

At the University of Kentucky Medical Center over 350 patients with gynaecological tumours were treated in the six years from 1977 (Maruyama *et al* 1983). The sources used contained 20–25 μg of californium-252 and had an active length of 15 mm and an overall length of the order of 25 mm. The clinical technique involved inserting stainless steel tandems (uterine tubes), rubber ovoids and plastic or stainless steel straight tubes which were then subsequently loaded with the californium tubes. Initially, the patients were treated by giving external radiation of between 40 and 55 Gy in five to six weeks with a 10 MV photon beam prior to intracavitary therapy. The early work (Tai and Maruyama 1978) set out to study the clinical problems associated with using neutron therapy and a trial was made on 20 patients in which the neutron-to-gamma dose ratio and the dose rate were calculated for the various reference points chosen within the pelvis, the overall dose being calculated using an RBE of six for the neutron component of the dose.

The points chosen were those suggested by Maruyama *et al* (1976) as reference points for dose surveillance in gynaecological treatments and are as shown in table 7.11. The results of this study showed that for points near the tandem, Points D and T, the neutron/gamma ratio was approximately 2.0, at Point A it was 1.85 and at B and C 1.25, as shown in the table. This finding led Tai and Maruyama (1978) to conclude that the favourable neutron/gamma ratio would suggest that californium-252 offered effective and potential biological advantages for intracavitary therapy.

Table 7.11 Neutron/gamma ratio at anatomical reference points.

Point	Anatomical reference description	Neutron/gamma ratio
A	Manchester system reference points.	1.85
B	Manchester system reference points.	1.26
C	Represents the external iliac nodes at the lateral pelvis.	1.25
D	The pelvic inlet, above the tandem applicator; represents dose to the small bowel.	2.28
T	The point 1 cm above the cervico-vaginal junction and 1 cm lateral to tandem.	2.02
V	Represents vaginal surface dose at vault and lateral surface of vaginal ovoid.	1.84
R	Represents the anterior wall of the rectum at level of the cervico-vaginal junction.	1.79
BL	Represents the posterior wall of the bladder at the trigone.	1.77

Following radiobiological considerations, Maruyama (1979) and Maruyama *et al* (1980a) reported a change in the dose schedule so that the intracavitary therapy was now to be given before or early in the external beam therapy treatment when there was the maximum quantity of bulky, necrotic and hypoxic tumour present. The schedule given was 150–200 neutron-cGy to Point A followed, by 50–55 Gy in five to six weeks of external beam therapy. In late 1980 the results of a series of 59 patients treated with this regime were published (Maruyama *et al* 1980b) in which the authors reported a significant improvement in local tumour control when compared with post external beam neutron brachytherapy and when compared with the conventional photon beam therapy. Indeed, the authors claimed an improvement in local control and clearance of tumours in approximately 90% of the patients treated, although they acknowledged the need for a longer follow-up time to determine long-term local control and late normal tissue reaction.

The rationale for using this regime stemmed from the fact that neutron therapy with high RBE presents greater damage to hypoxic bulky tumours than normal tissue. Following five years of clinical experience based on over 350 patients, a subsequent report (Maruyama *et al* 1983) stated that the use of low-dose-rate neutron brachytherapy produced five-year cures without a high frequency of normal tissue complications. This was attributed to the high relative RBE for bulky hypoxic tumours, which along with the OER advantage and the dose-rate-independent effects produced rapid tumour regression and good local control. In addition, adjacent normal tissues which are oxygenated have a lower RBE value than that of hypoxic tumour, thus californium-252 brachytherapy increases the dose differential, therapeutic gain and probability of local tumour control, since the high RBE of californium-252 for hypoxic tumour is much less in normal tissue. These differences favour tumour control and less normal tissue damage.

Techniques used for this form of brachytherapy were all after-loading ones to minimise the potential hazards to staff. Implant times never exceeded eight hours and the patients were treated in shielded rooms to which access was strictly controlled. Use was made of Perspex for the carrier to transport the sources to the treatment room and a 1.5 cm Perspex bed shield was used to provide staff protection. It is well recognised that radiation safety is an essential pre-requisite for the use of neutrons in brachytherapy and the problem is even more acute when gynaecological tumours are being treated. These problems have been considered by many not to be cost effective in terms of the benefits to be gained from using neutrons in comparison with photons, and this is one of the reasons why the use of californium has not gained wider popularity. To improve the situation a remote after-loading technique and treatment facility for use with californium-252 has been described by Onai *et al* (1978). In spite of these problems the protagonists for the use of neutrons in the treatment of gynaecological and pelvic cancer believe that these early results warrant continued use of californium-252 in these clinical conditions.

8 Practical Aspects of Absorbed Dose Calculations

8.1 Introduction

Calculation techniques for determining the absorbed dose in the vicinity of implanted sources are now well established. Numerous commercially available computer software packages now enable the user to produce dose distributions in many planes within the target volume, thereby giving the radiotherapist more detailed information than was feasible with manual calculation methods. The benefit of these advances in computational techniques, however, is limited by the accuracy with which the spatial position of the sources can be determined due to the high dose gradients associated with brachytherapy.

For certain superficial lesions, for example in breast implants where iridium wire is loaded in either plastic or rigid tubes, it is often possible to make direct measurements of the relative positions of the wires. These positional data can act as the basis of dose calculation. However, with other interstitial and intracavitary brachytherapy where the source positions cannot be seen externally, radiographic techniques must be applied to determine the source position. Initially, films were used to determine the actual volume implanted so that corrections could be made to the treatment time as well as enabling the radiotherapist to check that no two sources were so close together that they could cause an area of excess dose with the consequent problem of necrosis. Whilst the radiotherapist sets out to achieve an 'ideal' distribution based on the Paterson–Parker or Paris systems, for example, anatomical and technical difficulties often preclude the achievement of the ideal.

Visualisation techniques play an important role in brachytherapy. In addition to determining the volume of the implant, radiographic techniques have been used to define the relative spatial position of the sources. Many authors have described the use of radiographs to construct three-dimensional models of the source positions, either by using mechanical devices for rapid model construction (Mussel 1956a) or by using light sources shining onto a model of the implant which is then adjusted so that the shadows produced correspond to the distribution of the sources as seen on orthogonal films (Vaeth and Meurk 1963, Spira and Horn 1967). In this latter technique the light sources are placed at the effective focus of the x-ray beam, thereby ensuring the same divergence of the light beam as the x-ray beam. Spira and Horn (1967) extended this technique to determining dose distributions due to an implant by placing a suitable x-ray film within the volume and using hollow tubes to construct the model, which are then loaded with sources similar to those used in the implant. Numerous other devices and instruments have been described to aid the determination of the source positions and doses to selected points (Holt 1956, Mussel 1956b). All of these techniques, together with graphical methods for determining source position, are an attempt to improve the accuracy with which it is possible to calculate the absorbed dose due to sources implanted into a patient or inserted into one of the body cavities.

8.2 Radiographic Techniques for Source Localisation

8.2.1 Orthogonal films

In this radiographic technique, two films are taken at right-angles to each other, thereby allowing the position of the sources to be reconstructed. Consider two points P and Q (figure 8.1), the ends of a tube or implanted needle, with coordinates (x_1, y_1, z_1) and (x_2, y_2, z_2). The projection of these points will be P', Q' and P'', Q'' on the AP and lateral films respectively. The determination of the distance between the points is given by applying Pythagoras's theorem such that

$$(PQ)^2 = (x_1 - x_2)^2 + (y_1 - y_2)^2 + (z_1 - z_2)^2. \tag{8.1}$$

To determine the length, PQ, from the radiographs, it is necessary to define a base line common to both films. This line is often taken to be a line perpendicular to the intersection of the radiographs and is usually parallel to the edge of the films. By measuring the distances $P'Q'$ and $O''P''$ directly from the two films then

$$(PQ)^2 = (P'Q')^2 + (O''P'')^2. \tag{8.2}$$

The lengths measured on the films, however, will be magnified due to beam divergence, thus a magnification for each film is required in order that the true length can be determined. If M_{ap} and M_{lat} are the magnification factors then equation (8.2) becomes

$$(PQ)^2 = (P'Q'/M_{ap})^2 + (O''P''/M_{lat})^2. \tag{8.3}$$

For this technique to produce accurate results, it is essential that

(i) the two films are mutually perpendicular,
(ii) the central ray of the x-ray beam is perpendicular to the film,
(iii) the patient does not move between exposures.

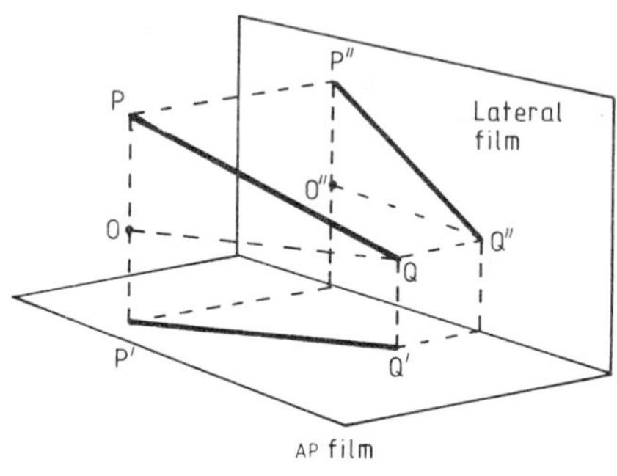

Figure 8.1 Diagram of projection of a tube or needle, PQ, onto antero-posterior (AP) and lateral films.

8.2.2 Shift films

The radiographic localisation of sources with this technique involves making a double exposure on a single film, the two exposures being taken with the x-ray tube in different positions, T_1 and T_2 separated by a distance S. Since the projection of the point $P(x, y, z)$ in figure 8.2 onto the film is $P_1(x_1, y_1, 0)$ from $T_1(0, 0, F)$ and $P_2(x_2, y_1, 0)$ from $T_2(S, 0, F)$(where F is the focal film distance OT_1)

$$x = x_1 S/(x_1 - x_2 + S) \tag{8.4}$$

$$y = y_1 S/(x_1 - x_2 + S) \tag{8.5}$$

$$z = F(x_1 - x_2)/(x_1 - x_2 + S). \tag{8.6}$$

Knowing the focal film distance, F (which equals OT_1), and $T_1T_2 = S$ and making measurements on a single film of the distances $(x_1 - x_2)$, x_1 and y_1, it is possible to localise the point P in space, providing the patient does not move during the two exposures.

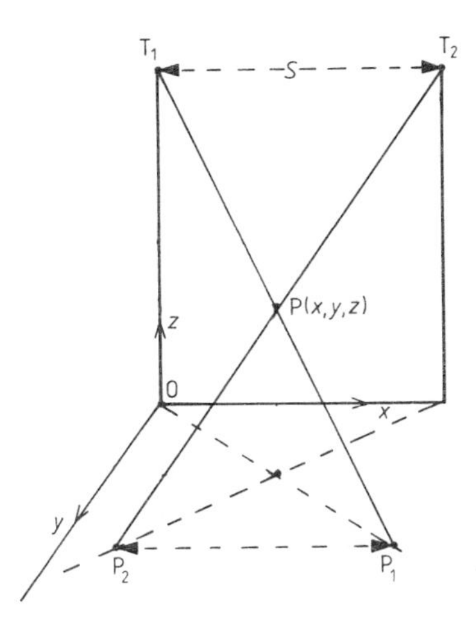

Figure 8.2 Principle of shift-film technique of taking double exposure on a film with the x-ray source at positions T_1 and T_2.

8.2.3 Stereo-shift films

An adaptation of the shift-film technique in which two separate films are taken of the sources, has been suggested by many workers (Nuttall and Spiers 1946, Mussel 1956a, Shalek and Stovall 1962). In order that the two films can be related to each other, a common reference point, such as a lead-shot marker, has to be projected onto both films. These projections can then act as the origin of the coordinate system for calculation. This situation is shown in figure 8.3 where the films are placed at a distance d from the plane containing the marker O. The projections of this marker onto film 1 and film 2 are O_1 and O_2 respectively. Similarly, a seed at $P(x, y, z)$ will appear at P_1 on film 1, taken with the tube at T_1, and at P_2 on film 2 after the tube has been moved in the y direction to T_2, a distance S. By similar triangles

$$O_2O_1/S = d/(F-d) \tag{8.7}$$

and

$$P_2P_1/S = z/(F-z) \tag{8.8}$$

where z is the distance of P above the plane of the films and F is the focal film distance.

If the distance O_1P_1 is y_1 and O_2P_2 is y_2, then, by solving equations (8.7) and (8.8) for z we obtain

$$z = F[(F-d)(y_1-y_2)+Sd]/[(F-d)(y_1-y_2)+SF]. \tag{8.9}$$

The magnification M of the seed P is given by $M = F/(F - z)$ and using equation (8.9) we can show that

$$M = (F - d)(y_1 - y_2) + SF/S(F - d)$$

thus $x = x_1/M$ and $y = y_1/M$. Knowing the distances F and S it is possible to determine the coordinates of all seeds, (x_i, y_i, z_i). By taking the origin of the implant relative to one of the seeds or ends of a tube, it is possible to construct the spatial distribution of the sources.

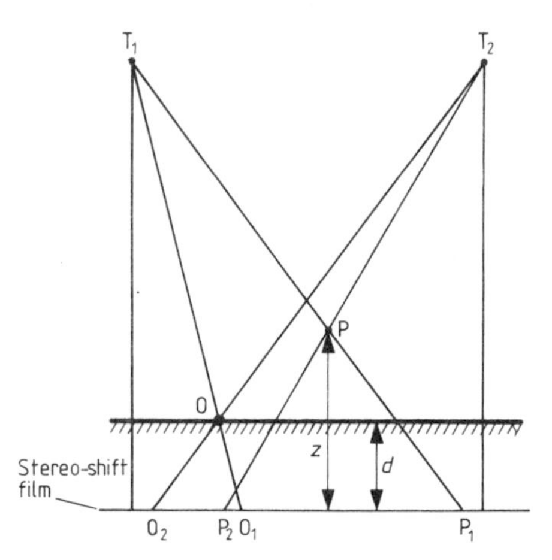

Figure 8.3 Stereo-shift film technique; single exposures made on two films which are then related by a common reference point O.

Another approach to taking stereo-shift films is to shift the patient a defined distance S_p, rather than the tube. It can be shown that in the above equation S has to be replaced by $-S_p$ (Shalek and Stovall 1969, Anderson 1975, Anderson *et al* 1981a, Mohan 1981). This technique requires the identification of the sources on both films, but as with orthogonal films it is essential that the position of the patient remains unchanged between the two exposures. Since shift films are usually carried out in the anterio-posterior direction of the patient, visualisation of the sources is often easier, particularly with seed implants in the pelvis where lateral films often result in poor definition of the sources.

8.2.4 Tomography and computed tomographic scans

To produce a distribution in the plane of the implant or insertion it is advantageous to produce a tomographic slice through the plane of the implant. Egan and Johnson (1960) described a technique for visualising radium sources in an implant using a multicassette unit. On each film the positions of the radium needles were traced as black dots and the films then positioned in a multiplane configuration corresponding to their positions in the multicassette. Metal wires, simulating the radium needles, were heated and pushed through the plastic sheets at the points indicated by the aligned black dots. The plastic sheets were fixed by the wires, thereby providing a three-dimensional model of the implant. A further tomographic approach for aiding the dosimetry of implants was described by Pierquin and Fayos (1962). In this technique the patient is positioned in such a way that an axial transverse tomogram is produced, thereby giving a transverse section of the body at the level of the implant. Immediately after the implantation of the sources, the position of the implant is determined by fluoroscopy, or taking orthogonal films, and the projection of the vertical axis of the implant is marked on the patient's skin surface. It is through this level that the tomographic sections are made. The images of the needles or wires appear as points on the film and these are used to make direct calculations of possible dose, rather than using an indirect method.

Techniques have also been developed for source localisation using a computed tomographic (CT) scanner. Wilkinson *et al* (1983) described such a localisation method for patients undergoing intracavitary therapy for carcinoma of the cervix using the Selectron after-loading system. These patients are scanned using dummy source trains about two hours after the insertion has been performed when the patient is more relaxed and cooperative. To aid localisation contrast media are introduced into the bladder and the rectal catheter is inflated with 20 ml of air. Scannograms are performed both in the AP and lateral directions, followed by a series of cross-sectional scans. Using this information, the positions of the source pellets within the applicators are found by setting up a coordinate system with the z axis parallel to the long axis of the patient and the cross-sectional scans in a series of x–y planes being defined by a single z value corresponding to the couch position. Once the sources have been localised it is possible to display the dose distribution on each slice due to the sources. This has the obvious benefit that it is then possible to relate the distribution to the anatomical structures visualised in the scan. Should the source arrangement result in vulnerable structures being over-irradiated, the sources within the applicators can be altered. Similar techniques can also be achieved for seed implants, although the limitations of the slice width of each scan make localisation in the sagittal plane difficult.

8.3 Reconstruction of Source Positions

8.3.1 Determination of magnification factors

To obtain the actual dimensions of sources or an implant from a radiograph, account must be taken of the magnification of the implant when viewed on the film. The following are possible methods of determining the magnification factor of an implant.

8.3.1.1 The geometric method. In this approach the magnification factor is obtained by calculation from a knowledge of the target film and implant film distances.

8.3.1.2 The direct needle method. This uses the known lengths of the sources to determine the magnification factor. For example, a needle shown on one film to be in a plane parallel to the plane of the other film is chosen since it enables the magnification of the implant to be determined by calculating the ratio of its measured length to its true length.

8.3.1.3 The indirect needle method. If a needle can be shown to be in the same plane and parallel to one of the dimensions to be determined, then not only is it magnified but also foreshortened to the same extent as the dimension. This method, however, is limited in use in practical terms to a regular single-plane implant.

8.3.1.4 The single ring method. In this technique a metal ring about 4 to 5 cm in diameter is placed on the patient. Since one of the diameters of a ring will always be magnified without foreshortening, it is possible to obtain the true magnification of the plane in which the ring lies by calculating the magnification factor. This is the ratio of the maximum projected diameter of the ring on the film to the actual dimension of the ring. Placing the single ring at the same distance from the source (as close as possible) as the implant is the basis of this approach, suggested by Meredith and Stephenson (1967). For a cervical application, for example, the ring is placed on the anterior surface of the patient over the sources when taking the lateral film and on the side of the patient at the level of the sources for the AP film. If the ring is at the same level as the sources, then the magnification factor for the ring on each film will be the same as the magnification factor for the implant. However, if it is not possible to place the single ring at the level of the sources then inaccuracies can occur in deriving the magnification factor of the implant unless correction is made. In the situation where the ring is displaced from the level of the implant (figure 8.4) then the magnification of the implant on the two films can be derived from the equations

$$M_{lat} = F/[(F/m_{lat}) \pm (x/m_{ap})] \tag{8.10}$$

$$M_{ap} = F/[(F/m_{ap}) \pm (y/m_{lat})] \tag{8.11}$$

where M_{lat} is the magnification of the implant on the lateral film, M_{ap} is the magnification of the implant on the AP film, F is the focus-to-film distances (T_1O' and T_2O''), m_{lat} is the magnification of the ring on the lateral film and m_{ap} is the magnification of the ring on the AP film. The term y is the distance between the position of the implant and the ring on the lateral film, R″S″, and x is the distance between the position of the implant and the ring on the AP film, R′S′.

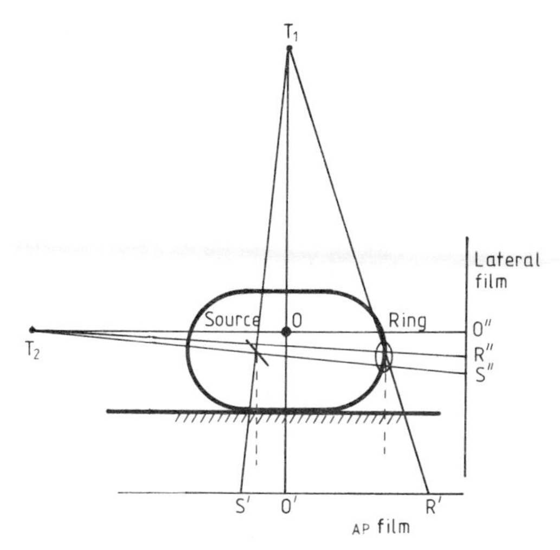

Figure 8.4 Use of a single ring for determining the magnification of an implant.

In equations (8.10) and (8.11) the plus sign is used when the implant is nearer the film than the ring; the negative sign being used for the converse situation. This technique provides the basis for calculating the true dimensions of an implant, as shown in the following example

Example 8.1

Consider films taken of a needle implant of a tongue, shown diagrammatically in figure 8.5. The ring magnification on the AP film is 1.45 whilst on the lateral it is 1.39, both films being taken with a focus-to-film distance, F, of 100 cm. The distances x and y on the two views are 2.80 and 7.40 cm respectively, thus using these values in equations (8.10) and (8.11) the magnification of the implant is given by

$$M_{ap} = 100/[(100/1.45) + (7.40/1.39)]$$
$$= 1.35$$
$$M_{lat} = 100/[(100/1.39) + (2.80/1.45)]$$
$$= 1.35.$$

The width of the implant measured on the lateral film is 4.20 cm, whilst the projection on the AP film measures 0.70 cm. Thus the true width of the implant is given by

$$[(4.20/1.35)^2 + (0.70/1.35)^2]^{1/2} = 3.15 \text{ cm}.$$

Knowing the height of the needles used, it is possible to calculate the true area of the implant, a requirement for the use of the Paterson–Parker tables for determining the treatment to give the prescribed dose.

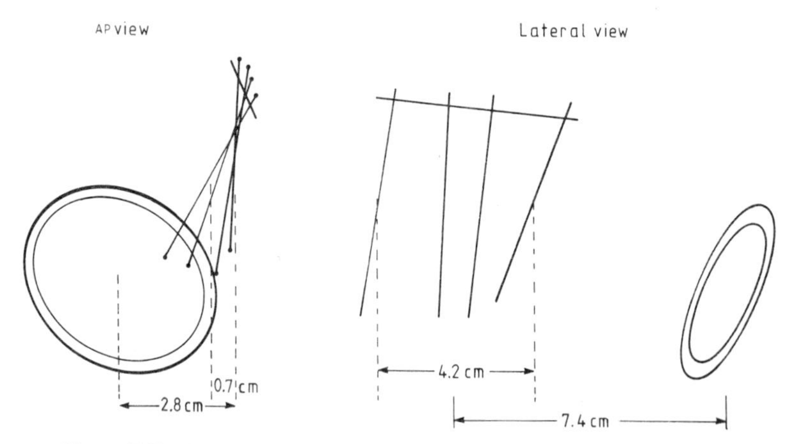

Figure 8.5 Diagrammatic representation of a needle implant of the tongue.

A further development of the single ring technique has been used by Mohan and Anderson (1974) in the brachytherapy computer software BRACHY. In this technique a ring, visible on both AP and lateral films, is used and by taking the coordinates of the end-points of the largest dimensions of the ring images on each film, the coordinates of the centre of the ring can be determined. This information is then used to calculate the ring magnification on the two films. Coordinates for the sources are also derived which, with the magnification factors for the source position derived from the ring magnification factors, are used to calculate the relative positions of the sources in the implant.

8.3.1.5 Multi-ring technique. To improve accuracy the single ring technique can be adapted for use with four rings, R_1, R_2, R_3 and R_4, one on each of the anterior, posterior and lateral surfaces of the patient. Provided all rings are visible on both films the magnification of any seed or needle in the implant can be obtained by taking the position of each point relative to the rings, using the magnification of the rings on the two films to determine the magnification of each needle. This approach obviates the need to know the focus-to-film distance, F, at which each film is taken. Considering the 'idealised patient' (figure 8.6) the x, z coordinates of any point, P, can be determined from the following relationships:

$$x/X_A = J - (z/F_1) \tag{8.12}$$

$$z/Z_L = K + (x/F_2) \tag{8.13}$$

where

$$J = \{(m_1A + m_2B)/[m_1m_2(A + B)]\}$$

$$K = \{(m_3C + m_4D)/[m_3m_4(C + D)]\}.$$

The terms m_1, m_2, m_3 and m_4 are the magnification of the rings R_1, R_2, R_3 and R_4. The other terms are, with reference to figure 8.6, $A = O''R_1''$, $B = O''R_2''$, $C = O''R_3'$, $D = O'R_4'$, $F_1 = T_1O'$, $F_2 = T_2O''$, $X_A = O'P'$ and $Z_L = O''P''$.

It can also be shown that the focus-to-film distances F_1 and F_2 can be derived from

$$F_1 = [(m_3C + m_4D)(A + B)m_1m_2]/[(C + D)m_3m_4(m_1 - m_2)] \tag{8.14}$$

$$F_2 = [(m_1A + m_2B)(C + D)m_3m_4]/[(A + B)m_1m_2(m_3 - m_4)]. \tag{8.15}$$

If only two rings are used, for example the rings on the anterior and posterior surfaces, then it can be shown that

$$z/Z_L = [F_1(m_1 - m_2)/(A + B)m_1m_2] + x/F_2. \tag{8.16}$$

In this situation we can see, by comparison with equation (8.13), that

$$K = F_1(m_1 - m_2)/(A + B)m_1m_2.$$

To a first approximation, if $z \ll F_1$ and $x \ll F_2$, i.e. the rings are directly over the implant, equations (8.12) and (8.13) become

$$x/X_A = J \quad \text{and} \quad z/Z_L = K.$$

The magnification factors of the implant therefore on the AP and lateral films are given by

$$M_{ap} = 1/J = m_1 m_2 (A + B)/(m_1 A + m_2 B) \tag{8.17}$$

and

$$M_{lat} = 1/K = m_3 m_4 (C + D)/(m_3 C + m_4 D) \tag{8.18}$$

where $x = X_A/M_{ap}$, $y = Y_A/M_{ap} = Y_L/M_{lat}$ and $z = Z_L/M_{lat}$.

The net effect of making this approximation can be seen in the following example by calculating values of x and z from known values of the parameters in figure 8.6.

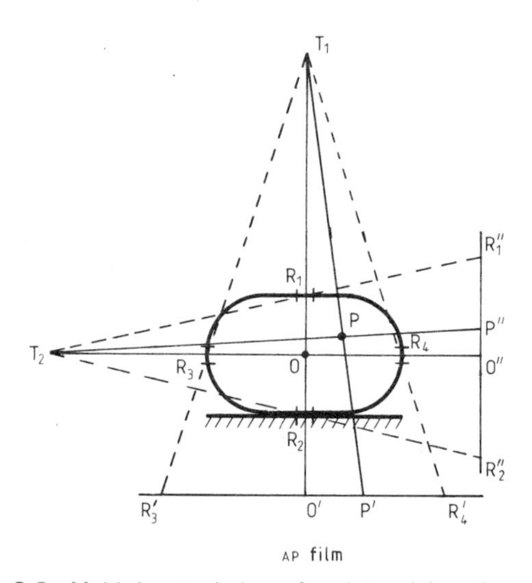

Figure 8.6 Multi-ring technique for determining the coordinates of source positions within a patient.

Example 8.2

Assume that films are taken of a seed at point P, where $x = z = 5$ cm, the focus-to-film distances, F_1 and F_2, in each case being 100 cm. The positions of the rings are such that $OR_1 = OR_2 = 10$ cm and $OR_3 = OR_4 = 15$ cm; also, the rings R_2 and R_4 are 5 cm from the AP and lateral films respectively. Using the above dimensions we can calculate that:

$A = \mathrm{O''R_1''} = 12.5$ cm, $B = \mathrm{O''R_2''} = 12.5$ cm, $C = \mathrm{O'R_3'} = 17.65$ cm, $D = \mathrm{O'R_4'} = 17.65$ cm. Also, $X_A = \mathrm{O'P'} = 6.25$ cm and $Z_L = \mathrm{O''P''} = 5.88$ cm. The rings R_1, R_2, R_3 and R_4 are magnified such that $m_1 = 1.33$, $m_2 = 1.05$, $m_3 = 1.54$, and $m_4 = 1.05$.

Using these calculated values in equations (8.14) and (8.15), we find that $F_1 = 99.86$ cm and $F_2 = 99.26$ cm, and that $J = 0.852$ and $K = 0.801$, which when used in equations (8.12) and (8.13) and solved for x and z gives $x = 5.01$ cm and $z = 5.01$ cm. If only the first approximation was made in the determination of the magnification factor of P then $x = 5.33$ cm and $z = 4.71$ cm, an error of $+3$ mm in x and -3 mm in the determined value of z. If P was the centre of an implant of area $6 \times 6\ \mathrm{cm}^2$ then this approximation would suggest an area of $6.4 \times 5.6\ \mathrm{cm}^2$. This error, however, only produces an error in the area of less than $0.2\ \mathrm{cm}^2$ which, when using the Paterson–Parker tables, will cause little error in dose. It must be emphasised, however, that this is one cause of error; other errors occur due to measurements made on orthogonal films.

An approximation often made in determining the magnification of an implant from rings is to use a method of simple proportions. From the lateral film it is possible to find the relative positions of the sources with respect to the anterior and posterior rings, $\mathrm{P''R_1''}$ and $\mathrm{P''R_2''}$. Similarly, on the AP film the distances $\mathrm{P'R_3'}$ and $\mathrm{P'R_4'}$ can be measured. The magnification of the sources on the AP film is then determined by calculating the difference in the magnification factors of the two AP rings and then adding or subtracting this amount to the appropriate ring magnification factor. A similar approach is adopted for the lateral film. Whilst this technique is not geometrically accurate, the errors obtained for separations of 20 cm between the rings for a focus-to-film distance of 100 cm and a film couch distance of 5 cm are only of the order of -1.5% maximum. For many clinical applications this degree of error is acceptable.

8.3.2 Errors in reconstruction techniques

Whilst the above radiographic techniques enable the reconstruction of source positions, there are inherent limitations which lead to inaccuracies in deriving the true source positions. These errors, which occur for both orthogonal and shift films, can lead to errors in computing the absorbed dose distribution and arise from:

(i) the accuracy with which measurements can be made on a film to determine the magnification factors;

(ii) the limitation of the calculational accuracy obtainable with the equations used for deriving source position;
(iii) the fact that the radiation beam may not be perpendicular to the films.

Errors in distances are particularly important in shift-film techniques. Hughes (1956) considered this problem in detail. If errors dx_1, dx_2 and dS are made in measuring x_1, x_2 and S, then the error in x has been shown by Hughes (1956) to be

$$dx = [(dx)^2_{x_1} + (dx)^2_{x_2} + (dx)^2_S]^{1/2}$$

where

$$(dx)_{x_1} = [S(S - x_2)/(S + x_1 - x_2)^2]\,dx_1$$
$$(dx)_{x_2} = [Sx_1/(S + x_1 - x_2)^2]\,dx_2$$
$$(dx)_S = [x_1(x_1 - x_2)/(S + x_1 - x_2)^2]\,dS.$$

A similar approach can be adopted to find the error in y. To determine the error in z, dz, it is convenient to let $(x_1 - x_2) = p$, the distance between the projections P_1 and P_2 which is directly measurable on the shift film. Equation (8.6) can be written in the form

$$z = Fp/(p + S).$$

If the errors involved in measuring p, F and S are dp, dF and dS, then by summing these errors in quadrature the total error in z can be determined from

$$dz = [(dz)^2_p + (dz)^2_S + (dz)^2_F]^{1/2}$$

where

$$(dz)_p = [FS/(p + S)^2]\,dp$$
$$(dz)_S = [-Fp/(p + S)^2]\,dS$$
$$(dz)_F = [p/(p + S)]\,dF.$$

The dominant factor in the error in z is that due to errors in measuring the distances p and S, since these are multiplied by the factor F. This error can be reduced by increasing the shift distance, S, or by decreasing the focus-to-film distance, although there are practical limitations to this imposed by the x-ray equipment. Fitzgerald and Mauderli (1975) reported a series of experiments designed to optimise the parameters involved in producing shift films and suggested that for $F > 100$ cm and $F/S < 2$, to achieve a sufficiently accurate three-dimensional construction of an implant, the tube shift should be of the order of 50–60 cm. The effect of these positional errors can be to yield different and incorrect isodose distributions.

Taking distance measurements from orthogonal films also results in errors which affect both the calculation of magnification factor as well as the calculation of the coordinates of the source positions. In the simplest case of a single ring placed at the level of the implant then with film measurement errors of ±0.5 mm the resulting error in the magnification factor on the AP and lateral films will be of the order of ±0.02. Such an error would result in an error of 0.1 cm^2 on the area of an implant as calculated in §8.3.1, with the consequent error in dose of 0.6%.

When equations (8.17) and (8.18) are used to determine magnification factors, the errors in measuring $O''R_1''$, A, and $O''R_2''$, B, and in determining the values of m_1 and m_2 result in an overall error for M_{ap} which is given by

$$(\mathrm{d}M)^2_{\mathrm{ap}} = (\mathrm{d}M)^2_{m_1} + (\mathrm{d}M)^2_{m_2} + (\mathrm{d}M)^2_A + (\mathrm{d}M)^2_B.$$

Similarly, if we use equations (8.12) to (8.16) to derive x and z, we know that errors will occur in determining the magnification factor as well as in making measurement on films and in establishing the value of F. Again, the resulting error in x and z can be determined by considering the error in each parameter and summing these in quadrature such that

$$(\mathrm{d}x)^2 = (\mathrm{d}x)^2_{X_A} + (\mathrm{d}x)^2_J + (\mathrm{d}x)^2_K + (\mathrm{d}x)^2_F + (\mathrm{d}x)^2_{Z_L}$$

and

$$(\mathrm{d}z)^2 = (\mathrm{d}z)^2_{Z_L} + (\mathrm{d}z)^2_K + (\mathrm{d}z)^2_F + (\mathrm{d}z)^2_x.$$

The major factor which contributes to the error in x is that which occurs in determining the magnification factor of the rings and hence the associated magnification factor of the source position. In both the two- and four-ring techniques the error in J remains the same, whereas the error in K is dependent upon the method used. If only two rings are used, as is the case with equation (8.16), then the error in K, $\mathrm{d}K$, due to m_1 and m_2 is given by

$$\mathrm{d}K_{m_1} = F\ \mathrm{d}m_1/[(A+B)m_1^2]$$
$$\mathrm{d}K_{m_2} = F\ \mathrm{d}m_2/[(A+B)m_2^2].$$

However, with a four-ring technique the error in K is determined from equation (8.13), that is

$$\mathrm{d}K_{m_3} = -D\ \mathrm{d}m_3/[(C+D)m_3^2]$$

and

$$\mathrm{d}K_{m_4} = -C\ \mathrm{d}m_4/[(C+D)m_4^2].$$

Since the value of $(C+D)$ is a factor of three to five less than the value of F, then the overall error in K is appreciably reduced.

In the analysis of the sources of error so far we have assumed that the central axis of the x-ray beam is normal to the films. This is not always the situation (see figure 8.7). If the x-ray tube is displaced from O to T_2, then the two points P_1 and P_2, which only differ in the x direction, will be projected to P_1'' and P_2'' respectively, whereas when the beam is normal to the film the images of P_1 and P_2 are superimposed. The consequence of such a tube displacement results in the following equations

$$s/s'' = (F-a)/a$$

$$s/(s''+t) = [F-(a+x)]/(a+x)$$

where $s = OT_2$, $s'' = O''P_1''$, $a = P_1O''$, $t = P_1''P_2''$ and $x = P_1P_2$. Therefore

$$t = sFx/[(F-a-x)(F-a)].$$

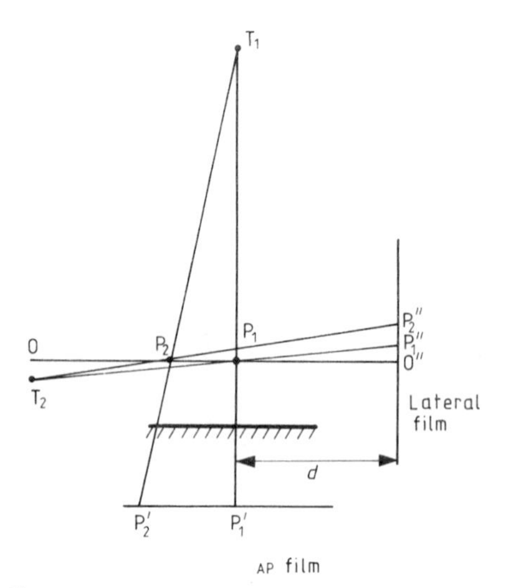

Figure 8.7 Errors occurring when the central axis of x-ray beam is not normal to the film.

The apparent z position of P_2 relative to P_1 is given by

$$z/t = (F - a)/F$$

thus

$$z = sx/(F - a - x).$$

It is conceivable that the lateral beam could be 2° from the horizontal, thus for $a = 20$ cm and $F = OO'' = 100$ cm then $s = 2.8$ cm. If x is 5 cm from the central axis of the AP beam, then the apparent z position of P_2 on the lateral film, P_2'', is 0.19 cm from O''. This error will be in addition to the errors in determining the x, y and z coordinates of a point within a patient or phantom.

8.3.3 Comparison of stereo-shift and orthogonal radiography

The high dose gradient around implanted sources means that any error in determining the source position results in error in calculating the absorbed dose distribution. The magnitude of these errors differs with the radiographic technique adopted. Although these spatial errors rarely result in errors in absorbed dose of greater than $\pm 5\%$ around the implant or gynaecological insertion, appreciable errors can occur when brachytherapy and external beam therapy are combined due to errors in determining source positions in relation to patient anatomy.

To illustrate these inherent inaccuracies in the different radiographic techniques we will consider irradiating a rectangular phantom of cross section 20×30 cm^2 at the centre of which is a small metallic sphere, P. On each of the surfaces are placed rings of 5 cm diameter and the origin of the coordinate system, O, is such that the point P has coordinates (5,5). For each film the focus-to-film distance = 100 cm and the phantom surface nearest the source is at 75 and 65 cm respectively, which means that the film is 5 cm beyond the phantom in each case. We also assume that all films are taken with the beam normal to both the film and the phantom. For orthogonal films (figure 8.8(*a*)) the beam is centred over the ring on the anterior surface R_1 for the AP film and over R_3 for the lateral film. For the stereo-shift films (figure 8.8(*b*)) the first image is obtained by centring the tube over R_1 and the second image obtained after the x-ray tube has been moved from T_1 to T_2, a distance S. Using three dimensions for the phantom, we can calculate values for the parameters X_A, Z_L, A, B, C and D on the orthogonal films, X_A, x_s, W and V for the shift films and the magnified ring sizes $m_{R_1'}$ and $m_{R_2'}$ on the AP film and stereo-shift films and $m_{R_3''}$ and $m_{R_4''}$ on the lateral film. These

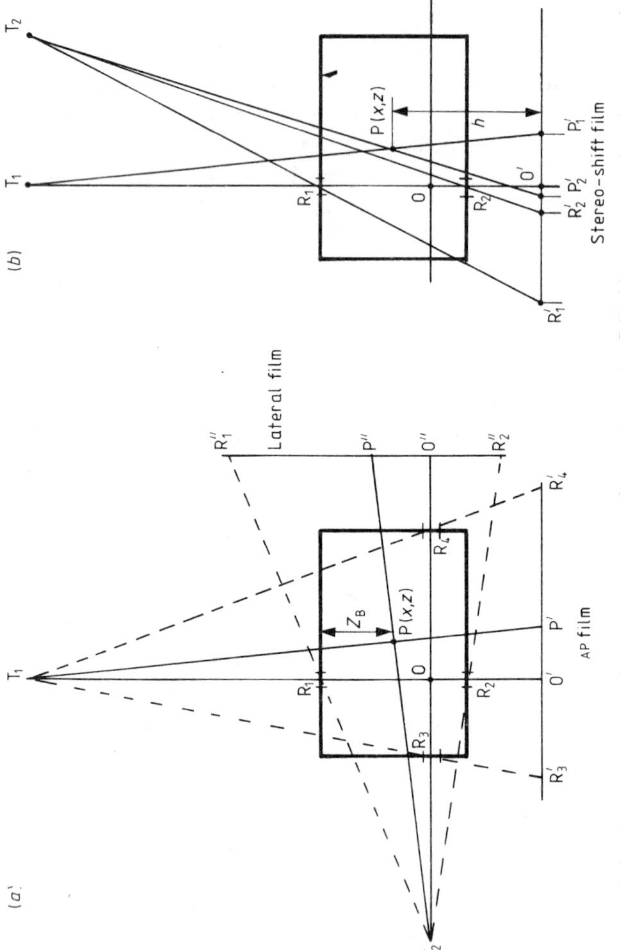

Figure 8.8 Determination of the coordinates of a seed at point P inside a rectangular phantom using (*a*) orthogonal film technique, (*b*) stereo-shift film technique.

values, shown in table 8.1, have been derived numerically from the dimensions of the phantom and rounded to the nearest 0.5 mm to simulate typical measurements obtained in practice with a ruler. The error associated with each measurement on a film is taken to be ± 0.05 cm and that in determining F and S ± 0.5 cm. Using these values in equations (8.12) and (8.13) we can derive the parameters for the two- and four-ring techniques, together with their associated errors which are obtained by summing the errors in quadrature.

Table 8.1 Values of various parameters for different films.†

AP film (values in cm)	Lateral film (values in cm)	Shift film (values in cm)
$X_A = 5.90 \pm 0.05$	$Z_L = 6.25 \pm 0.05$	$X_A = 5.90 \pm 0.05$
$m_{R'_1} = 6.65 \pm 0.05$	$m_{R''_3} = 7.70 \pm 0.05$	$m_{R'_1} = 6.65 \pm 0.05$
$m_{R'_2} = 5.25 \pm 0.05$	$m_{R''_4} = 5.25 \pm 0.05$	$x_s = 3.55 \pm 0.05$
$C = 11.10 \pm 0.05$	$A = 20.00 \pm 0.05$	$W = 6.65 \pm 0.05$
$D = 22.20 \pm 0.05$	$B = 6.65 \pm 0.05$	$V = 5.25 \pm 0.05$
$C + D = 33.35 \pm 0.05$	$A + B = 26.65 \pm 0.05$	$S = 20.0 \pm 0.50$
$F = 100.0 \pm 0.5$	$F = 100.0 \pm 0.5$	$F = 100.0 \pm 0.50$

† Parameters refer to figures 8.8(*a*) and 8.8(*b*): $X_A = O'P'$, $Z_L = O''P''$, $F = T_1O' = T_2O''$, $A = O''R''_1$, $B = O''R''_2$, $C = O'R'_3$, $D = O'R'_4$, $W = R'_1R''_2$, $V = O'R'_2$ and $x_s = P'_1P'_2$.

In the case of shift films we can see that

$$S/x_s = (F - h)/h \qquad (8.19)$$

therefore

$$h/F = x_s/(S + x_s).$$

Also

$$x/X_A = (F - h)/F \qquad (8.20)$$

therefore

$$x = X_A S/(S + x_s).$$

Using the rings it is not necessary to know the value of S and the distance from the posterior ring to the film, $O'R_2$, can be calculated, thereby enabling the position of P to be calculated with reference to the posterior surface of the phantom. Thus

$$S/W = [F - (t_a + d)]/(t_a + d)$$
$$S/V = (F - d)/d.$$

Also

$$m_{R_1'} = F/[F - (t_a + d)]$$

and

$$m_{R_2'} = F/(F - d)$$

where $t_a = R_1R_2$ and $d = O'R_2$. Therefore

$$(t_a + d) = F[1 - (1/m_{R_1'})] \tag{8.21}$$

and

$$S = W/(m_{R_1'} - 1). \tag{8.22}$$

Using the values shown in table 8.1 in the relevant equations above we can calculate the value of x and Z_B where $Z_B = (t_a + d) - h$, and the associated error in this process. The results are given in table 8.2 for both orthogonal-film and shift-film techniques.

Table 8.2 Results for various parameters.

Orthogonal films		Shift films	
Two rings	Four rings	Using derived value for S	Using measured value for S
$J = 0.902 \pm 0.010$	$J = 0.902 \pm 0.010$	—	—
$K = 0.752 \pm 0.063$	$K = 0.749 \pm 0.007$	—	—
$x = 5.03 \pm 0.08$ cm	$x = 5.03 \pm 0.07$ cm	$x = 5.02 \pm 0.06$ cm	$x = 5.01 \pm 0.05$ cm
$z = 5.02 \pm 0.40$ cm	$z = 5.00 \pm 0.06$ cm	—	—
$Z_B = 10.02 \pm 1.32$ cm	$Z_B = 9.98 \pm 0.15$ cm	$Z_B = 9.83 \pm 1.19$ cm	$Z_B = 9.74 \pm 1.04$ cm

It is apparent that the method which achieves the coordinates of P, (5, 5) with respect to the origin or (5, 10) with respect to the anterior surface of the patient, with the least possible error is the four-ring technique, particularly with reference to the accuracy with which the z coordinate can be determined. However, in much of the matching of external to intracavitary therapy we are interested in the accuracy of the x coordinate; here there is little to choose between the various methods.

Various authors have carried out practical studies using phantoms to show the relative accuracy of orthogonal and stereo-radiographic techniques. Sharma *et al* (1979, 1982) evaluated the performance of the two radiographic options offered by a commercially available computer implant dosimetry program (Artronix PC12). These programs allow data to be entered from orthogonal or stereo radiographs taken with either a single-plane fiducial jig or biplane fiducial jig. The performance of all systems was evaluated by their ability to determine the length of the sources used and the angles between the sources. The ability to determine length was carried out using sources 4 cm and 8 cm long set on a styrafoam phantom such that for each length one source was parallel and the other perpendicular to the direction of the shift; the shift films being taken at a focus-to-film distance of 100 cm and with a shift of 20 cm. For each option the sources were sampled six times and the lengths calculated, together with the standard deviation on the derived lengths. The results indicated that the orthogonal option showed less variation in the calculated values.

Whilst this and the biplane stereo technique determined the length correctly, the single-plane fiducial option gave an error of 1 cm in determining the 4 cm length. When considering the ability of the system to determine the angle between two sources, it was found that whilst the orthogonal technique gave the correct angle, the calculated length was 1.5 mm in error, whereas the biplane stereo option gave an error of 4° in angle and 2.5 mm in length and the single-plane option gave the correct length but 11° error in angle. These results confirmed that the orthogonal option provides the more accurate reconstruction technique. In later experiments Sharma *et al* (1982) repeated the work using a simulated interstitial seed implant and again found that the random measurement errors on the reconstructed coordinates from stereo films were three to five times that of the orthogonal films. The authors did point out that whilst the relative individual seed positions may be in error by as much as 1 cm, the errors do not significantly change the shape or area of the isodose curves surrounding the implant periphery.

Whilst these positional errors are significant with the shift-film techniques, they can be minimised by making the focus-to-film distance greater than 100 cm and increasing the shift S to a value greater than 60 cm as mentioned above, although this is often not feasible in clinical practice. Since, however it is not possible to visualise seeds adequately on a lateral film, stereo techniques are advantageous in such cases without causing clinically significant variation in the dose distribution.

8.3.4 Aids to improved radiographic reconstruction

Techniques to improve the accuracy of determining source positions have developed over the years. Many early techniques used mechanical aids to improve localisation (Holt 1956, Mussel 1956b), whilst Vaeth and Meurk (1963) developed the Rotterdam radium reconstruction device. In 1967 Hidalgo *et al* described a precision reconstruction system which used a frame placed over the patient. This frame contained lead markers, in the shape of a cross, on all four sides and allowed two films to be placed at right-angles to each other. The images of the frame markers and sources on the two films, togther with the knowledge of the relative positions of the films to the markers on the frame, a and b, can then be used to determine the origin of the cartesian coordinate system for the calculation of the coordinates of each source.

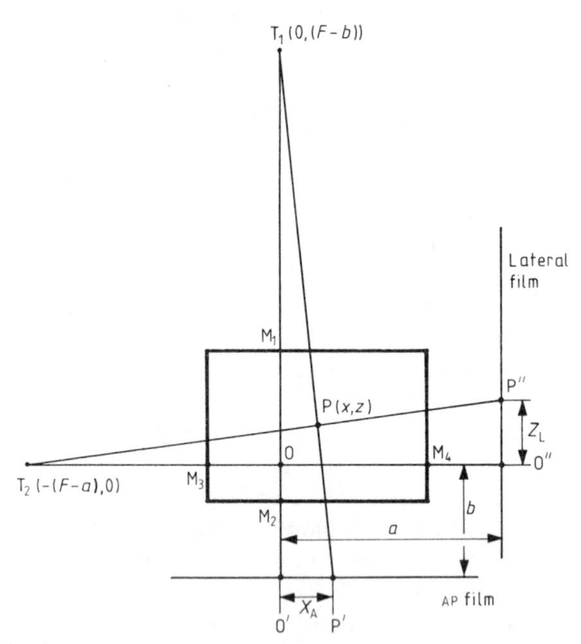

Figure 8.9 Use of lead-shot markers in a 'bridge' as an aid to determining source coordinates.

Consider therefore the situation, shown in figure 8.9, of an object $P(x, z)$ being radiographed from two source positions T_1 and T_2 to produce two orthogonal films. If the x-ray beam from T_1 is centred over the marker M_1 and is normal to the planes containing the markers M_1 and M_2, then the image of M_1 will be superimposed on M_2. Similarly, the image of M_3 will be superimposed on M_4 if T_2 is centred over M_3 and the beam is normal to the planes containing M_3 and M_4. For this configuration it is possible, using data taken from the films, to derive straight-line equations which describe the paths of rays passing through P which emanate from the focal spots T_1 and T_2. These equations are

$$z = -(Fx/X_A) + (F - b) \tag{8.23}$$

for AP film and

$$z = Z_L[x - (F - a)]/F \tag{8.24}$$

for lateral film.

These two equations can be solved to determine x and z by making the measurements of X_A and Z_L from the film and by taking measurements of the projection of the 'marker cross' on the film. By comparing them with the true dimensions of the cross, it is possible to determine the value of F, rather then relying on an accurate knowledge of that dimension.

A similar approach to deriving positional information has been developed by Rosenwald (1970). In both methods the value of the y coordinate of point P is obtained by using the magnification of the object on the AP film, M_{ap}, determined from

$$M_{ap} = x/X_A = [(F - b)F - Z_L(F - a)]/(F^2 - X_A Z_L).$$

These techniques have obvious advantages for computer calculation methods, since it is not necessary to determine magnification factors separately. In cases where orthogonal films are not always appropriate, such as with seed implants, other methods have been suggested. To produce stereo-radiographs suitable for determining source position, Shalek and Stovall (1962) proposed the use of special equipment which produced two planes of fiducial markers, as indicated in §8.2.3. This consists of a plate beneath the patient in which are embedded eight pieces of lead shot which delineate a $20 \times 20\ cm^2$ square. Attached to this plate is a plastic arm which contains four pieces of lead shot, originally positioned in a plane parallel to the plate, with a separation of 30 cm between the two sets of lead shot. The images of the 12 shot are projected onto the stereo-radiographs and constitute 'fiducial points'. From the 'shot' positions on the film it is possible to compute the tube shift and the target film distance; again it is not necessary to make allowance for magnification of the images.

Another method of stereo x-ray photogrammetry using a stereo x-ray comparator has been described by Van Kleffens and Star (1979) and Storchi and Van Kleffens (1979). The equipment consists of two x-ray tubes and a pneumatically driven cassette charger, thereby obviating the need to move either the x-ray tube or patient in producing stereo-radiographs. This approach involves taking three films, two for the stereo and one for an antero-posterior picture. Viewing of the stereo film is carried out on a specially constructed comparator, the films being positioned by correctly aligning the images of the 'markers' which are produced by a grid placed above the film cassette. The authors claim that this technique improves the accuracy of determining the coordinates of the sources, thereby improving the dosimetry around sources.

The use of radiotherapy treatment planning simulators for source visualisation has also been recommended by many authors. These units have the distinct advantage of being manufactured to a high degree of accuracy, thus the many parameters required for accurate source localisation are known to greater accuracy than can be determined by the use of diagnostic x-ray units, even when they are used with rings, fiducial jigs or bridges. It can be shown that since the gantry angle is known to within $\pm 1^\circ$ and the focus-to-film distance to within ± 5 mm, the accuracy in determining source position can be improved by a factor of two.

8.4 Multisource Localisation and Dose Computation Algorithms

8.4.1 Dose computation algorithms

One of the first references to the use of computers in interstitial implant dose calculations was by Nelson and Meurk (1958) who described a system of summing the contribution to any point from any three-dimensional distribution of point sources by using the inverse square law relationship. Over the intervening years many authors (Shalek and Stovall 1961, Stovall and Shalek 1962, Powers *et al* 1965, Rose *et al* 1966, 1967, Jameson *et al* 1968, Dale 1976) have made numerous improvements and produced more elaborate programs for interstitial dose calculation for needles and seeds which incorporate the various other factors mentioned in Chapter 3. The computer calculations of dose distribution around intracavitary sources have also been numerously described (Hope *et al* 1964, Snelling *et al* 1965, Jameson *et al* 1968,

Batten and Shalek 1966, Batten 1968, Vickery and Redpath 1977, Mohan and Anderson 1974, 1978, 1981, Anderson 1975) whilst in 1972 Stovall and Shalek published a review of computer techniques then available for interstitial and intracavitary radiotherapy (Stovall and Shalek 1972).

In these various techniques the dose distributions around the sources are produced for various planes and in a variety of orientations. Once the source positions within the patient are determined radiographically, many systems compute the total distribution from first principles each time, by summing the contribution each segment of each source contributes to a point in a 'patient' grid, whilst others use stored data related to the sources used and sum the contribution of each source to the grid points. Hope *et al* (1964) calculated dose by segmenting a line source of radium into five points and deriving the dose rate to grid points by integrating the Sievert integral using Simpson's rule. The resulting grid data were then used to produce an isodose distribution. Instead of computing the dose afresh at each point, Snelling *et al* (1965) stored data at 800–900 points for each type of source used. A similar approach was adopted by Rose *et al* (1966, 1967) who calculated the dose rates at an array of points in a plane, defining the points in terms of polar coordinates with the centre of the container as the origin and defining the dose at millimetre intervals from 0 to 10 mm and at angular intervals of one degree from 0 to 90°, i.e. a total of 101×91 points in one quadrant. Again the dose rates were computed by applying Simpson's rule in 1° increments to the form of the Sievert integral appropriate for the zones of interest defined in §3.6.2. Combined dose distributions in any chosen plane are then computed using the data stored on magnetic tape, the distance from the origin to the point in the patient and reference grids for each source used being calculated by coordinate geometry. Various other methods of data storage have been used, such as the dose tables of Breitman (1974) for caesium-137 tubes and needles.

In the BRACHY II computer program developed at the Memorial Hospital, New York, in 1981 the data stored are either the specific dose rate factor, F_D, if the source is specified in terms of its activity, or the relative dose rate factors, R, if specified in terms of source strength as described in §3.5. The dose to any point within the patient reference grid is then calculated using either equation (3.20) or (3.22) above. The advantage of adopting the relative dose rate factor is that it is a dimensionless quantity. This eliminates the dominant geometry factor and thus is a slowly varying function of position. As a consequence, the amount of data required to compute brachytherapy dose distributions is considerably reduced. Obviously, prior to any dose computation for clinical brachytherapy, the data tables have to be derived and entered into the program. Whilst these techniques represent various methods of calculating dose to points or dose distributions, it is necessary to provide manual or computer algorithms for determining the source positions within the patient.

8.4.2 Computer algorithms for determination of needle and tube positions

The positional coordinates of each source can be obtained using the techniques previously described (§8.3.4) and the use of computers to determine these coordinates is well documented. Many authors have described techniques for the use of digitising methods to enter source data from either orthogonal or stereo-radiography (Adams and Meurk 1964, Rose *et al* 1966, Stovall and Shalek 1968, Shaw and Thomas 1973, Mohan and Anderson 1974, 1978, 1981). This information is then used to compute the dose distribution around the sources. Others, however, carried out checks on the calculated coordinates and derived lengths of sources in an attempt to reduce the errors associated with radiographic techniques.

The input data used by Hope *et al* (1964) were the measurements of the ends of each tube from stereo-radiographs. Once calculated the program adjusted the coordinates of the ends of the tube if the calculated tube length was incorrect so that in dose computation the known lengths of the standard tubes were used. In an attempt to reduce errors, Batten (1968) described a method of determining the relative positions of the stereo-radiographs and x-ray tube direct from the projection of the fiducial points on the radiographs and corrected any errors due to film shift or orientation by minimising the 'square errors' of the parameters which led to the errors. Another reconstruction algorithm using a bridge 'marker' system was described by Donati and Volterrani (1973). They derived a set of equations to determine the coordinates of the x-ray positions from two orthogonal films by making measurements of the position of the reference markers. These equations $f_i (i = 1, 6)$ were then solved to achieve a minimum of the function F, where $F = \Sigma f_i^2$. Knowing the coordinates of the x-ray source, six further equations were derived for the radioactive source positions and again solved by a similar process.

Vickery and Redpath (1977) gave details of a reconstruction algorithm which automatically located needle positions from two orthogonal radiographs. For this technique orthogonal films are taken on an x-ray simulator and the source positions digitised. These provide the input for the needle-matching program. Since the radiographs are taken on a simulator, it is possible to choose the axis of rotation of simulator as the axis common to both films. The program requires that the tip and butt of each needle are digitised in that order, although the order of needles on each film need not be the same. After completing the digitising process the program continues to match needles in two films by adding the square of the difference in the x coordinates of the tips to the square of the differences in the butt coordinates for any needle in film 1 and any needle in film 2. The smallest value of this function is used to make a match between the two films and the process is repeated until all needles are matched. Once the film coordinates have been determined, it is then possible to derive the true coordinates and the resultant reconstruction is displayed on a storage oscilloscope. To determine the treatment plane the operator can observe the rotation of the implant in any one of these dimensions and thus choose the appropriate plane of calculation. The ability to produce an interactive display which can be rotated about the x, y and z axes has also been described by Jameson (1975). This technique involved deriving data from shift films taken at 100 cm full-scale deflection and with a shift of 40 cm. This gave a reproducibility of source position of 1 or 2 mm, although no attempt was made to match needles automatically from the films, the operator being responsible for identifying the individual sources in each film.

8.4.3 Computer algorithms for determining positions of seeds

The identification of a large number of implanted radioactive seeds is extremely difficult, if not impossible. The possibility of using stereoradiographs for this purpose was discussed by: (a) Rose *et al* (1967), who used a scanning procedure on two accurately taken shift radiographs with a common reference line shown in both films; (b) Stovall and Shalek (1968), using the fiducial marker equipment; and (c) Mohan and Anderson (1974, 1978, 1981) in the BRACHY II group of programs.

In these techniques 'manual sorting' of the seeds is required. To overcome this limitation Amols and Rosen (1981) developed a computer algorithm which computes the location of the seeds from a set of two stereo- and one AP radiograph using a ray-tracing technique. The implant radiographs are taken in a simulator with the film cassette placed at the isocentre of the machine. The centre of the x-ray field is therefore the same point in space for all three films, which are taken at the same source axis distance with the gantry rotated at $+8°, 0°$ and $-8°$. The (x, y) coordinates of each seed are entered into the computer by means of a sonic digitiser together with the field centre on each film. The data obtained are then used to calculate the seed coordinates by projecting lines from the x-ray source to the film images of the seeds on both stereo films and noting the intersection point. This method alone can lead to incorrect seed localisation and thus the authors use the AP film to construct a third projection line. From the two stereo films it is possible to determine the (x, y) coordinates with an error of say 1–4 mm by matching the x_i and y_i coordinates. The AP film is then used to provide the additional criteria for the analysis. Since all three films are taken with the same x-ray source at the same y value only, seed images with the same y value on all three films can be paired together. To speed the matching process various bands of y values are chosen and the order acceptable for a match has been determined to be 1–4 mm. The authors claim that, whilst it is not possible to necessarily locate all seeds, provided 90% of the seeds are located there is no appreciable error in dose distribution.

In a subsequent paper by Rosen *et al* (1982) it was established that the tumour dose was accurate to within ±5% if all the seeds were reconstructed to within 5 mm of their actual position and, furthermore, up to 5% of the seeds could be mismatched between films with position error as larger as 20 cm without changing the tumour dose by more than 5%.

An adaptation of this three-film technique has been described by Rosenthal and Nath (1983). The three films are taken on a simulator, but this time the isocentre of the simulator is set in the centre of the implant at the source axis distance of 100 cm. Seed coordinates from the three isocentric radiographs, an anterio-posterior film and an orthogonal pair of films taken at geometry angles of $\pm 45°$ from the AP direction, are digitised and recorded in random sequence. From this the three-dimensional coordinates of the seeds in the implant are computed by matching the values of the common coordinates along the axis of rotation. To aid reconstruction two to five seeds are identified on each film to act as 'calibration seeds'. This rotate and match technique, the authors claim, leads to an accurate and consistent identification and reconstruction of seeds with 90% of them being within 2 mm of their actual location, thereby producing a small error in the calculated dose.

To overcome the possible inherent ambiguity when several seeds are coplanar with two radiographic source positions, Altschuler *et al* (1983) described a three-film technique whose x-ray source positions are in a plane which does not intersect the implant and are widely separated to achieve adequate parallax. The films are taken at simulator gantry angles of 310°, 30° and 0°, but with the couch translated 5 to 10 cm in the longitudinal direction for one of the geometry positions. All triangulation is done in an anatomical coordinate reference frame attached to the patient so that patient motion between dating the radiographs does not introduce errors. Again, a ray-tracing technique is employed in deriving coordinates and the intersection part is determined by the least square fit method.

8.4.4 Construction algorithms for implanted wires

The use of flexible after-loading wires, whilst being beneficial to both the radiotherapist and patient, can present difficulties in dose calculation since the dose at any point due to a wire is a function of its shape and its geometrical relationship to the point. Calculating dose in such implants has been carried out with various degrees of sophistication. Boisserie and Marinello (1979) described a method of calculating the mean basal dose and dose distribution in the central plane of the implant using a small Hewlett Packard calculator with associated printer. In the program the cartesian coordinates of the points of intersection of the wires with the central plane are entered manually into the program and are used to automatically determine the points for calculating the basal dose rate, the calculation of the basal dose and reference dose and finally the production of a distribution around the sources in the plane of interest. Whilst the approach uses the computer to carry out the dosimetry calculations of the Paris system, it does rely on the use of tomographic cross sections through the central plane or other manual methods of locating the sources in the plane and pays little or no attention to possible curvature of the sources.

Rosenwald (1975) described the use of a computer to automatically locate curved wires obtained in implants. The use of conventional AP and lateral films presents difficulties in determining points on both radiographs and, even if identification can be made, a point located on the AP projection cannot generally be identified on the lateral film because of the distortions of the actual shape of the wires. The author therefore derived a program, COUREP, which automatically correlates the corresponding wire and the points located on both radiographs. This technique is applicable to any iridium implant and can act as the basis of dose computation in three dimensions if each point on the wire is segmented and an integration technique is applied. To solve the problem of intersection of a defined line with the curved wire projection on both films, the shape and position of a point on the wire are described by a mathematical function containing Tchebycheff polynomials and the coordinates describing the wire are calculated by a least square method. To match common points $(x(t), y(t)), (y(u), z(u))$ on the two projections of wire, translation of coordinates is carried out and the y coordinate for a given value of t on one film is used to determine the y coordinate on the second by a method of continual approximation from one initial value of u.

The accuracy of this technique, as with others which require the taking of two consecutive radiographs, depends on there being no patient movement, but in routine work a displacement of 2–3 mm is acceptable. When entering the data for the reconstruction, the images of the wires are digitised, there being no particular rule for the choice of points except that both ends of each wire have to be selected and the number and sites of the points may be different on both films.

The use of orthogonal films taken on a simulator for the localisation of the implanted radioactive wires has also been described by Gantchew and Rosenwald (1976). In this technique a specially constructed twin light-box is used to determine corresponding points in the projection of the wire in the AP and lateral radiographs. The curved wires are divided into a series of straight segments and the coordinates of the centre of the segments on each radiograph determined. In order that unnacceptable errors do not occur, the points in the two films have to be chosen so that the actual segments of wire do not diverge too much from straight lines. As a check, the actual length of wire is calculated and the total length calculated must be less than 0.1 times the actual length. This approach has the advantage that non-uniformly active wires can be divided into segments and the activity of each segment can be considered separately in dose calculations. In the program the computation of the activity of each segment is determined from the total activity of the wire, this being possible for both uniformly and non-uniformly activated wires. Such an algorithm can then be included in the dose computation program to determine the distribution of dose around an implant.

9 The Safe Use of Sealed Sources

9.1 Protection Requirements

Whilst ionising radiation is used to kill maligant cells in brachytherapy, it can equally well kill or damage normal tissue both in the patient and in personnel involved in the treatment of the patient. It is therefore essential that any possible hazard to staff is minimised. With sealed sources the principle hazard comes not from contamination but from the emitted radiation. Protective measures must therefore be used to keep the dose to staff and members of the general public to levels as low as reasonably achievable, the ALARA principle. Dose limits have been set for all categories of personnel and are detailed in such reports as *Recommendations of the International Commission on Radiological Protection* (ICRP 1977, 1978). These limits are based on studies of the effects of radiation on

(i) people who are accidentally exposed to radiation,
(ii) survivors of the atomic bombs used in Japan in 1945,
(iii) animal experiments,
(iv) certain categories of patient treated with radiation.

The limits have been set such that non-stochastic effects in personnel are negligible and the risks of stochastic effects are reduced to a level where the risks are no greater than those of normal life.

To reduce exposure to a low level safe radiological working practices must be implemented in terms of source custody and handling. In addition, a well designed working environment incorporating adequate protective barriers must be provided. Due to the different energies of the gamma-ray photons emitted from the radionuclides used in brachytherapy, the protection afforded by absorbing materials differs, as can be seen in table 9.1.

Of the radionuclides typically used radium-226, cobalt-60 and caesium-137 require a greater thickness of absorber than the others,

Table 9.1 Half- and tenth-value layers in lead, steel and concrete for radionuclides used in brachytherapy. (Data taken from NCRP (1976).)

	Half-value layer (mm)			Tenth-value layer (mm)		
Nuclide	Lead	Steel	Concrete	Lead	Steel	Concrete
Cobalt-60	12	21	62	45	87	220
Iodine-125	0.025	—	—	—	—	—
Caesium-137	6.5	16	48	22	69	175
Iridium-192	6.0	13	43	16	43	147
Gold-198	6.0	12	41	11	41	135
Radium-226	16	22	69	45	87	234

particularly iodine-125. Whilst the use of protective barriers in preparation areas is of paramount importance, adequate protection must also be afforded to all staff during the transportation of sources within the hospital and during patient treatment. To reduce the dose to nursing staff bed shields are often used (figure 9.1). Horwitz et al (1964) showed the improvement in the radiation exposure rates which could be achieved by using 60 mg Ra eq caesium-137, rather than 60 mg of radium-226 both with and without a bed shield.

From these considerations it is apparent that the use of caesium-137

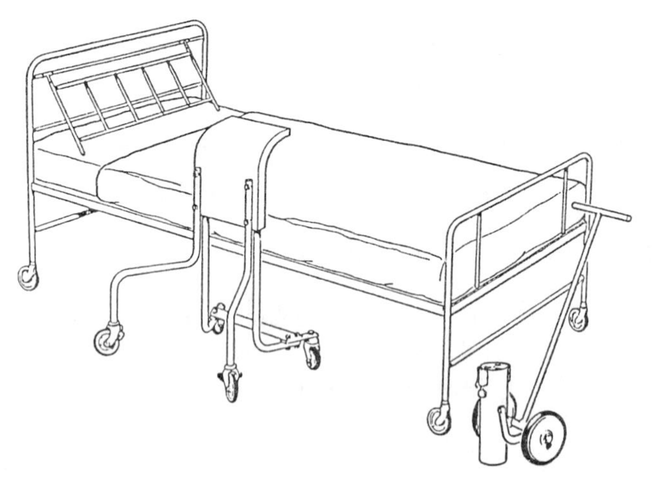

Figure 9.1 Movable lead bed shield, source transport trolley and shielded carrying container. (Reproduced by kind permission of Amersham International plc.)

for intracavitary therapy has many advantages from the radiation protection point of view. Obviously the dose levels around patients are dependent on the source activities used, thus with medium-dose-rate after-loading systems the levels will be higher and high-dose-rate after-loading systems require the patients to be treated in specially designed rooms. Glasgow *et al* (1985) derived the following relationship which gives the exposure rate, $\dot{X}(r)$, at 1 m in $\mathrm{mR\,h^{-1}}$ for an activity of A_c mg Ra eq of caesium:

$$\dot{X}(r)/A_c = (1.03 \pm 0.5)\exp[-(0.07 \pm 0.02)r]$$

where $2r$ is the pelvic diameter of the patient.

The use of iodine-125 also has distinct protection advantages, as demonstrated by Hilaris *et al* (1975). Measurements around patients undergoing treatment with this radionuclide, compared with treatment using radon seeds, give approximately ten times less exposure to staff and to members of the general public (Germain 1975). For prostatic implants Liu and Edwards (1979) have measured exposures of the order of $0.2\ \mathrm{mR\,h^{-1}}$ per mCi at 0.75 m from the patient.

When small sealed sources are used the UK Code of Practice, 'Code of Practice for the Protection of Persons against Ionising Radiation from Medical and Dental Use (1972)' and the 'Ionising Radiation Regulations 1985' give stringent guide-lines for the use of these sources. These codes recommend the following procedures.

(i) A separate room should be used, marked with the internationally accepted radiation symbol, for the preparation of the sources. The room must be well ventilated and eating, smoking and drinking are prohibited in the room.
(ii) Suitable shields should be provided to ensure adequate protection for the eyes when manipulating beta-ray sources.
(iii) Local rules and schemes of work for the protection of people in the proximity of the patient undergoing sealed-source therapy should be established. These rules should include

   (a) a statement that a minimum distance of 2.5 m must be kept between bed centres if patients have to be treated in a general ward;
   (b) instruction that the bed of a patient undergoing treatment must be identified with the standard radiation symbol, together with details of the nursing procedure to be adopted if the exposure rate at 1 m exceeds $50\ \mu\mathrm{Sv\,h^{-1}}$ ($5\ \mathrm{mR\,h^{-1}}$);

(c) the rule that no patient with sources in or on their bodies should be permitted to leave the ward or treatment room without the approval of the appropriate medical officer;
(d) a statement that nursing staff should not remain in the vicinity of the patient being treated any longer than necessary

In formulating these rules, the radiological protection supervisor and the sealed-source custodian must consider all the relevant information available about source preparation and nursing procedures in the light of the recommended codes of practice.

9.2 Storage and Custody of Sources

The potential hazard associated with the loss of a sealed source requires that sources can be located within a department at any given time. The responsibility for ensuring adequate safety precautions is that of the sealed-source custodian. To effect this degree of safety the codes of practice insist that there must be a register of stock of all sealed sources having a half-life greater than a few days. This register must contain detailed specifications of each individual source, its identification marks and serial number, together with reports on leakage tests and repairs. In addition, when sources are not in use they must be stored in a lockable, protected safe to which access is restricted. The construction of the safe depends upon the radionuclides and their form. For manual-loading gynaecological sources safes are often used which consist of drawers into which a known number of sources can be placed (figure 9.2). This storage method, however, is not suitable for storing iridium-192 wire which can be used in more than one patient. For this situation a lockable safe of the type shown in figure 9.3 can be used to store both wire and hair-pins.

These safes are manufactured of lead and/or steel to provide the necessary radiation protection and the safes themselves are located in the specially designed, well ventilated sealed-source preparation room. The siting of the safe within the room must be such that there is good accessibility between the safe and the working surface where the sources are prepared for clinical use. Whenever the safe or the preparation room are not being used they should be kept locked. With remote after-loading systems the treatment units themselves often act as the safe, as is the case with the Selectron where the design ensures that the maximum exposure rate on the surface does not exceed 25 $\mu Sv\,h^{-1}$ (2.5 $mR\,h^{-1}$).

Figure 9.2 Sealed-source safe showing drawer assembly for storing individual sources.

Whenever a source or set of sources is removed from the safe an issue sheet is prepared which remains with the sources until their return to the safe. Each time the sources or set of sources are transferred from one responsible person to another a signature must be obtained. When the sources are no longer required for use they must be returned to the main safe and ideally not left in an intermediate transit safe. To ensure the total integrity of the stock a periodic safe count must be made and the result noted in the sealed-source register. As sources will of necessity move from the sealed-source preparation room, it is advantageous to have a visual display indication of the location of all the sources within the building. In those clinical situations where a patient is treated by a permanent implant a record must be kept of the administration of the sources.

Whilst these safety precautions are designed to prevent the loss of a source, sources can be mislaid in transit or lost from a patient. In the event of such a suspected loss each department must have prepared in advance a routine which must be followed. In outline this policy should be as follows.

(i) All avenues by which sources could leave the last known vicinity should be immediately sealed. There must be no movement of staff, patients, laundry, dirty dressings or rubbish of any sort from the area and no toilets or sluices should be flushed.

(ii) The radiation protection supervisor for the area and the sealed-source custodian should be notified immediately.
(iii) A search must be initiated using competent personnel with appropriate monitoring equipment.

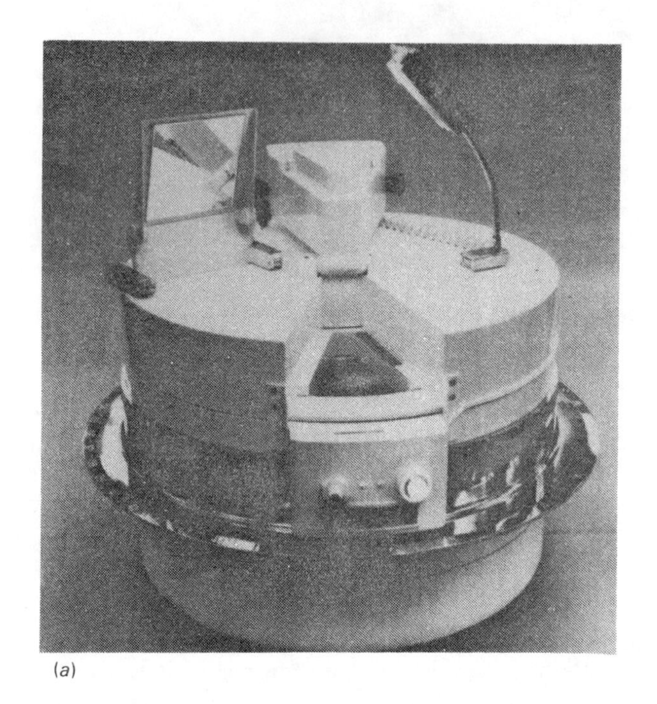

(*a*)

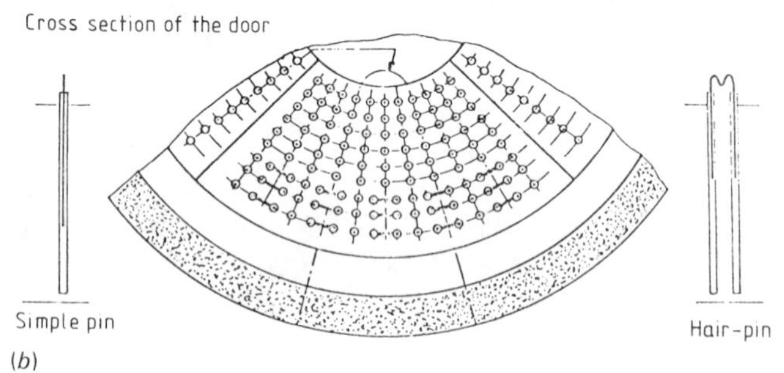

(*b*)

Figure 9.3 (*a*) Shielded storage safe for iridium-192 sources. (*b*) Top view through open door of iridium-192 storage safe. (Reproduced from Pierquin *et al* (1978a) by kind permission of Warren Green.)

(iv) If the initial search fails to locate the missing source then it may be necessary to ask for the assistance of specialists such as the National Radiological Protection Board (in the UK).

To minimise the chance of such a loss, it is advantageous to have permanent gamma-ray alarms on the exits to the areas where sealed sources are routinely used; these alarms require routine checking. It is good radiological safety practice to monitor all patients who have received sealed-source therapy prior to discharge. This is particularly important following the removal of after-loaded iridium wire by inexperienced staff since it is possible for the wire to be cut or for the tube to be removed without the wire. Similarly, all rubbish bags, dirty dressings and bedpans should be monitored before disposal.

9.3 Routine Source Checks

Prior to delivery the sources will have undergone tests in accordance with the British Standard BS5288 (1976) which is in agreement with the International Organisation for Standardization (ISO) publications ISO 1677, *Sealed Radioactive Sources—General*, ISO 2919, *Sealed Radioactive Sources—Classification*, and ISO/TR 4826–1979(E) *Technical Report, Sealed Radioactive Sources—Leak Test Methods*. These reports detail standard methods for testing sealed sources, including the following.

(i) Wipe test: a source is wiped with a swab moistened with ethanol or water and any activity is measured to ensure that it is less than the acceptable limit of 185 Bq (0.005 μCi).

(ii) Bubble test: the source is immersed in a suitable liquid such as ethanediol and the pressure in the vessel reduced to 100 mm of mercury. In this test no bubbles must be observed.

(iii) Immersion test L: the source is immersed in water or other suitable liquid at 50 °C for four hours and the activity in the water measured to ensure that the level is less than 185 Bq (0.005 μCi).

(iv) Immersion test M: the source is immersed in water which is raised to 100 °C and held at that temperature for ten minutes. The water is removed, the source cooled and the procedure repeated twice. The final water is measured to ensure that the activity is less than 185 Bq (0.005 μCi).

In the case of alpha-particle-emitting isotopes such as radium and radon the source is immersed in a solution of an organic liquid phosphor under vacuum. Any leakage of radon can be measured by liquid scintillation counting, the acceptable limits of detectable activity being 5×10^{-11} Ci per 25 hours. Sources produced by neutron activation also have the problem that the outer cladding remains slightly radioactive, thus they are not classified as a 'sealed radiation source'. However, in their design the outer wall material is chosen to minimise this possibility and any residual activity has been shown to be negligible. In addition to these radiation checks, all sources undergo other safety checks to ensure their suitability for use in brachytherapy.

Whilst manufacturers carry out numerous checks on sources prior to dispatch, including those mentioned above, it is desirable to repeat a wipe test on each source on delivery and also to carry out an autoradiograph on sources in which the radionuclide is mixed with a filler. One technique for achieving this is to place the source on a film for a few seconds, whilst at the same time giving it a short exposure with low-kilovolt x-rays. Checks for leakage must also be carried out annually by performing a simple wipe test. If between checks there is any cause to suspect damage to the source then it must be carefully examined and if necessary returned to the manufacturer for repair.

These precautions are particularly important for sealed radium sources, since any leakage of radon is extremely hazardous. Methods of testing for radon leakage can be carried out as mentioned above. However, if these facilities are not available, groups of about five or six needles or tubes can be placed, overnight, in test-tubes with cotton wool, the test-tubes being sealed for that time. After 24 hours the cotton wool is monitored with a Geiger–Müller counter for signs of activity, which is indicative of a leaky source. To establish which of the sources is faulty, each one is placed into a separate test-tube and the procedure repeated. The results of all tests on sealed sources, together with any repairs which take place must be recorded in the sealed-source register.

9.4 Source Handling and Transport

Preparing sources for use in brachytherapy can be a hazardous procedure if it is done without adequate protection and safety measures. All source handling must be done using appropriate tools such as forceps or, if the workload justifies it, slave manipulators, but on no account must sources ever be handled. When needles or tubes are to be used it is customary to thread coloured silks through the eyelets. These have a twofold purpose, namely to aid removal from the patient and secondly to provide a means of checking that all the sources are still *in situ* during treatment. In the threading process the sources must be held in a lead block so that only the eyelet protrudes.

All handling processes must be carried out with a protective barrier between sources and staff. These barriers, usually of lead, take various forms (figure 9.4) and enable radiation to the trunk and abdomen to be reduced. However, consideration must also be given to scattered radiation from other surfaces, thus it is advisable that the bench itself should have some protection material incorporated in it to reduce the photon flux towards the floor. Using this type of protected preparation bench behind which the operator stands, only the arms, hands and head are exposed to radiation, but even then the exposure to the head and eyes can be reduced by providing a protected observation window. In case any of the sources develops a leak, it is advisable to finish all surfaces with a plastic laminate.

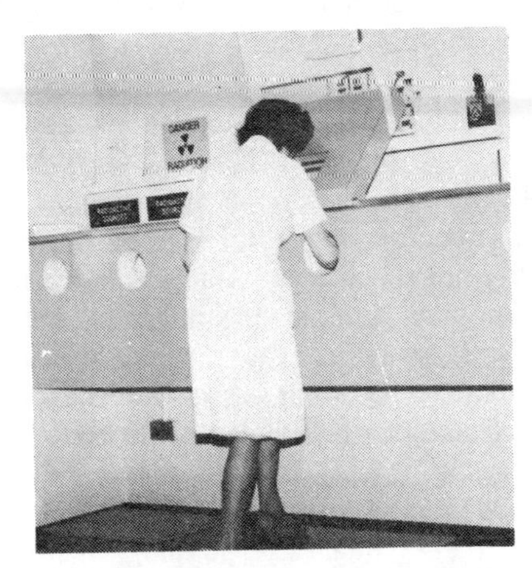

Figure 9.4 Protected lead barrier in sealed-source preparation room.

Protective barriers alone, however, will not be totally effective unless the operator is skilled and works swiftly and accurately when manipulating sources. Time and distance in the vicinity of the sources are equally important and in deciding safe working practices the time–distance relationship must always be considered. Little advantage is to be gained by insisting that an operator uses 30 cm long forceps to thread a silk in a needle, for example, if it takes three times longer than using 20 cm long forceps. Another simple guide-line to aid reducing the exposure dose to staff is to include a statement in the local rules and scheme of work that the sources should only be removed from the safe in limited numbers and that all other sources should be in protected areas during the preparation time. This same philosophy should also be adopted when sending sources to the operating theatre for use; it is not good radiation practice to have all the sources in theatre throughout the total time of the operative procedure, unless adequate protection can be provided.

To enable the easy transfer of sources to theatre just prior to their requirement, it is advisable to have the sealed-source preparation area adjacent to the operating theatre if possible. The sterilisation and subsequent cleaning of sources should only be carried out in protected areas. Sterilisation can be achieved by a variety of methods such as cold sterilisation liquids, ethylene oxide gas, or by boiling in water (although this method must be carried out with great care to ensure that the system cannot boil dry). This is particularly important when radium is used. To avoid the possibility of causing a source to fracture, the temperature of the sterilising unit must not exceed 180 °C. Whichever method is chosen adequate protection must be afforded to staff working in the sterilising area.

On return to the sealed-source preparation room after use the sources must be cleaned. Immersion in saline or hydrogen peroxide solution immediately after removal will ensure that any blood does not harden on the surfaces of the source, if this is not done then staff will need to spend excessive time trying to remove such deposits with the consequent increase in exposure dose. Cleaning can also be achieved by using a low-power ultrasonic cleaning bath, adequately protected with lead. The use of certain after-loading systems means that it is only necessary to clean the applicators used.

In the case of the manual after-loading of iridium wire, safety precautions have to be adopted in preparing the wire for insertion into the patient. Various manufacturers have developed equipment for loading the iridium wire into the polythene tubing. These loaders (figure 9.5) are designed to work either behind the normal protective barrier or to work on a normal laboratory bench, since they themselves have been built with adequate protection. In both cases the wires are loaded into the polythene to the desired length and secured in position by a heat or crimp seal immediately behind the wire. Once encapsulated, the sources are ready to insert into the outer polythene tubes inserted into the patient at the time of operation. With this technique, care must be exercised on removal to ensure that the wire is not cut and possibly left in the patient. Iridium wire hair-pins, as supplied, are often too long and require cutting before use. This must also be done with great care since fragments of iridium can easily be left on the cutting tool. To aid cutting Collins (1975) has designed a cutter which not only trims the hair-pins to the correct length but also provides a receptacle for the unwanted wire and any fragments produced in the cutting.

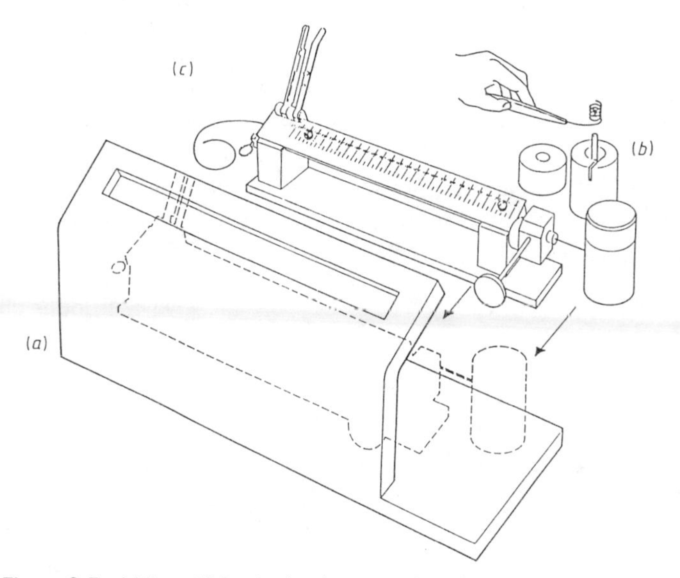

Figure 9.5 Iridium-192 wire loader assembly showing: (*a*) the protective shield comprising 3 cm of lead and a lead-glass window, 6 cm thick; (*b*) storage pot which enables wire to be fed directly into loader through a slot in the side of the pot and (*c*) the loader for loading 0.3 mm diameter iridium-192 wire into the inner nylon tubing. (Reproduced by kind permission of Amersham International plc.)

Of necessity, sources have to be transported within a hospital, either in a patient or between the treatment area and the sealed-source preparation room. In all cases the route taken must be the shortest possible and if the sources are within a patient then the accompanying staff must be advised on the best way to move the patient to minimise the dose to themselves. The movement of sources prior to and after use should only be done in specially designed containers or trolleys (see figure 9.1). In the treatment area the implanted sources still present a hazard and schemes of work must be formulated for nursing staff and any possible visitors. Whilst bed shields can reduce the exposure in certain directions, radiation levels in other positions are important and thus nursing staff must be advised on the optimum way to approach the patient to minimise the dose. Obviously, in all this patient care is of paramount importance, thus contingency plans must be drawn up to cope with the situation of caring for a patient who requires intensive nursing until the clinical decision has been made on whether or not to remove the sources. Patient surveillance using close circuit television can be of great benefit in reducing nursing staff dose. Many of these precautions, however, are minimised or even unnecessary when remote after-loading systems are used.

The safe custody of sealed sources is, therefore, most important. Administrative procedures must be rigorously adhered to so that all source movements can be traced, and local rules and schemes of work designed to ensure that the dose levels received by staff are as low as reasonably achievable commensurate with cost.

References

Adams G D and Meurk M L 1964 *Phys. Med. Biol.* **9** 533
Allt W E C and Hunt J W 1963 *Radiology* **80** 581
Altschuler M D, Findlay P A and Epperson R D 1983 *Phys. Med. Biol.* **28** 1305
Amols H and Rosen I I 1981 *Med. Phys.* **8** 210
Anderson L L 1975 in *Handbook of Interstitial Brachytherapy* ed B S Hilaris (Acton, MA: Publishing Sciences Group) p 87
—— 1976 *Med. Phys.* **3** 48
—— 1983 in Advances in Radiation Therapy Treatment Planning *Medical Physics Monograph No* 9 ed A E Wright and A L Boyer (New York: American Institute of Physics) p 542
Anderson L L and Ding I-Y 1975 in *Afterloading: 20 Years of experience* ed B S Hilaris (New York: Memorial Sloan-Kettering Cancer Center) p 149
Anderson L L, Kuan H M and Ding I-Y 1981a in *Modern Interstitial and Intracavitary Radiation Cancer Management* ed F W George III (New York: Masson) p 9
Anderson L L, Wagner L K and Schauer T H 1981b in *Modern Interstitial and Intracavitary Radiation Cancer Management* ed F W George III (New York: Masson) p 1
Andrews J R 1968 *The Radiobiology of Human Cancer Radiotherapy* (Philadelphia: W B Saunders)
Attix F H 1980 *Med. Phys.* **7** 254
Bateman T J, Davy T J and Skeggs D B L 1983 *Br. J. Radiol.* **56** 401
Batho H F and Young M E J 1964 *Br. J. Radiol.* **37** 689
—— 1967 *Br. J. Radiol.* **40** 785
Batten G W Jr 1968 *Am. J. Roentgenol.* **102** 673
Batten G W Jr and Shalek R J 1966 in Computer Calculation of Dose Distributions in Radiotherapy *Technical Report Series No* 57 (Vienna: International Atomic Energy Agency) p 100
Bednorel D R, Lanzl L H and Rozenfeld M 1976 *Med. Phys.* **3** 403
Belcher E H and Geilinger J E 1957 *Br. J. Radiol.* **30** 103
Berger M L 1968 MIRD Pamphlet No 2 *J. Nucl. Med.* **9** *Suppl. No* 1 17
Berkley L W, Hanson W F and Shalek R J 1981 in Recent Advances in Brachytherapy Physics *Medical Physics Monograph No* 7 ed D R Shearer (New York: American Institute of Physics) p 38
Berry R J and Cohen A B 1962 *Br. J. Radiol.* **35** 489
Berry R J, Paine C H, Wiernik G, Stedeford J B H, Barber C, Weatherburn H and Young C M A 1979 in Californium-252 Radiobiology and Medical Applications *Int. Symp. on Californium-252 Utilisation (Brussels) 1976* vol 1 ed D J Mewissen (SC du Pont de Nemurs) p 22
Bier R, Small R C, Leake D L and Kelley D M 1973 *Radiology* **108** 711
Biggs P J and Kelley D M 1983 *Med. Phys.* **10** 701

Biggs P J, Wang C C and Gitterman M M 1986 *Int. J. Radiat. Oncol. Biol. Phys.* **12** 247
Binks C 1981 *Br. J. Radiol.* **54** 269
Bloch P, Krishnaswamy V and Hale J 1972 *Am. J. Roentgenol.* **115** 822
Bloedorn F G 1956 *Am. J. Roentgenol.* **75** 457
Blomfield G W 1961 *Br. J. Radiol.* **34** 755
Boag J 1966 in *Radiation Dosimetry* vol II ed F H Attix and W C Roesch (New York: Academic) p 2
Boisserie G and Marinello G 1979 *J. de Radiol.* **60** 327
Boyer A L, Cobb P D, Kase K R and Chen D T S 1981 in Recent Advances in Brachytherapy Physics *Medical Physics Monograph No* 7 ed D R Shearer (New York: American Institute of Physics) p 22
Breitman K 1974 *Br. J. Radiol.* **47** 657
British Commission on Radiological Units (BCRU) 1984 *Br. J. Radiol.* **57** 941
Bryant T H E 1966 *Br. J. Radiol.* **39** 291
Buchan R C T and Griffin T C 1969 *Br. J. Radiol.* **42** 312
Busch M 1968 in *Role of Computers in Radiotherapy* (Vienna: IAEA) p 93
Cameron J R, Suntharalingham N and Kenny G N 1968 *Thermoluminescent Dosimetry* (Madison, Wis: University of Wisconsin Press) p 182
Campbell E M and Douglas M A 1966 *Br. J. Radiol.* **39** 537
Casebow M P 1971 *Br. J. Radiol.* **44** 618
—— 1984 *Br. J. Radiol.* **57** 515
—— 1985 *Br. J. Radiol.* **58** 549
Cassell K J 1983 *Br. J. Radiol.* **56** 113
Castro J R, Oliver G D, Withers H R and Almond P R 1973 *Am. J. Roentgenol.* **117** 182
Chan B, Rotman M and Randall G J 1972 *Radiology* **103** 705
Cobb P D and Bjarngard B E 1974 *Am. J. Roentgenol.* **120** 211
Cobb P D, Chen T S and Kase K R 1981 *Int. J. Radiat. Oncol. Biol. Phys.* **7** 259–62
Cohen L 1950 *Br. J. Radiol.* **23** 25
—— 1968 *Br. J. Radiol.* **41** 522
Cole A, Moore E B and Shalek R J 1953 *Nucleonics* **11** 46
Cole M 1963 in *The Treatment of Malignant Disease by Radiotherapy* ed R Paterson 2nd edn (London: E Arnold) p 331
Collins L 1975 *Br. J. Radiol.* **48** 49
Colvett R D, Rossi H H, Krishnaswamy V 1972 *Phys. Med. Biol.* **17** 356
Comité Français pour la Mesure des Rayonnements Ionisants 1983 *Recommendations pour la Determination des Doses Absorbées en Curietherapie Rapport CFMRI No* 1 (Paris: Bureau National Metrologie)
Corner G A, Kirk J and Perry A M 1982 *Clin. Radiol.* **33** 145
Corner G A, Skinner D L and Watson E R 1982 *Clin. Radiol.* **33** 477
Cowell M A C and Laurie J 1967 *Br. J. Radiol.* **40** 43
Cunningham D E, Stryker J A, Velkley D E and Chung K 1981 *Int. J. Radiat. Oncol. Biol. Phys.* **7** 121
Dale J W G, Perry W G and Pulfer R F 1961 *Int. J. Appl. Radiat. Isotopes* **10** 65
Dale J W G and Williams A 1964 *Int. J. Appl. Radiat. Isotopes* **15** 567
Dale R G 1976 *Br. J. Radiol.* **49** 533
—— 1982 *Br. J. Radiol.* **55** 748
—— 1983 *Med. Phys.* **10** 176
—— 1985 *Br. J. Radiol.* **58** 515

—— 1986 *Med. Phys.* **13** 963
Darby J, Lasbury B and Paine C H 1972 *Br. J. Radiol.* **45** 778
Debierne A and Regaud C 1915 *C.R. Acad. Sci., Paris* **161** 422
Delclos L 1982 in *Topical Reviews in Radiotherapy and Oncology* vol 2 ed T J Deely (Bristol: J Wright) Ch 7 p 175
—— 1984 in *Technological Basis of Radiation Therapy: Practical Clinical Applications* ed S H Levitt and N du Tapley (Philadelphia: Lea and Febiger) Ch 11
Delclos L, Fletcher G H, Sampiere V and Grant W H III 1978 *Cancer* **40** 970
Delclos L, Fletcher G H, Suit H D, Sampiere V and Moore B 1970 *Radiology* **96** 667
Diffey B L and Klevenhagen S C 1975 *Phys. Med. Biol.* **20** 446
Dobbie M D 1953 *J. Obstet. Gynaecol. Br. Commonw.* **60** 702
Donati G and Volterrani F 1973 *Acta Radiol.* **12** 268
Doss L L and Richman M S 1979 *Br. J. Radiol.* **52** 810
Dudley R A 1966 in *Radiation Dosimetry* vol II ed F H Attix and W C Roesch (New York: Academic) p 326
Durrance F Y and Fletcher G H 1968 *Radiology* **91** 140
Dutreix A 1974 *J. Radiol. Electrol.* **55** 781
Dutreix A, Marinello G, Pierquin B, Chassagne D J and Houlard J P 1979 *J. Radiol. Electrol.* **60** 319
Dutreix A, Marinello G and Wambersie A 1982 *Dosimetrie en Curietherapie* (Paris: Masson)
Dutreix A and Wambersie A 1968 *Acta Radiol.* **7** 389
—— 1975 *Br. J. Radiol.* **48** 1034
Dutreix J 1961 *J. de Radiol.* **42** 228
Easson E C and Pointon R C S 1985 *The Radiotherapy of Malignant Disease* ed E C Easson and R C S Pointon (New York: Springer)
Egan R and Johnson G C 1960 *Radiology* **74** 407
Eldridge J S and Crowther P 1964 *Nucleonics* **22** 56
Ellis F 1963 *Br. J. Radiol.* **36** 153
—— 1968 in *Current Topics in Radiation Research* **4** 359 (Amsterdam: North-Holland)
—— 1985 *Int. J. Radiat. Oncol. Biol. Phys.* **11** 1685
Ellis F and Sorensen A 1974 *Radiology* **110** 681
Ellis F and Taylor C B G 1982 *The Amersham Caesium-137 Afterloading System of Gynaecological Brachytherapy* (Amersham: Amersham International plc)
Ernst E C 1949 *Radiology* **52** 46
Evans R D 1955 *The Atomic Nucleus* (New York: McGraw-Hill)
—— 1968 in *Radiation Dosimetry* vol 1 ed F H Attix and W C Roesch 2nd edn (New York: Academic) Ch 3 p 94
Evans R D and Evans R O 1948 *Rev. Mod. Phys.* **20** 305
Eve A S 1906 *Phil. Mag.* **12** 189
Failla G 1937 *Radiology* **29** 202
Fitzgerald L T and Mauderli W 1975 *Radiology* **115** 455
Fletcher G H 1953 *Radiology* **60** 77
—— 1980 *Textbook of Radiotherapy* 3rd edn (Philadelphia: Lea and Febiger)
Fletcher G H, Shalek R J, Wall J A and Bloedorn F G 1952 *Am J. Roentgenol.* **68** 935
Fletcher G H, Wall J A, Bloedorn F G, Shalek R J and Wootton P 1953 *Am. J. Roentgenol.* **61** 885
Fletcher G H, Wootton P, Storey W H and Shalek R J 1954 *Am. J. Roentgenol.* **71** 1021

Fowler J F 1963 *Phys. Med. Biol.* **8** 1
—— 1966 in *Radiation Dosimetry* vol II ed F H Attix and W C Roesch (New York: Academic) p 291
Fowler J F and Attix F H 1966 in *Radiation Dosimetry* vol II ed F H Attix and W C Roesch (New York: Academic) p 241
Fox M and Nias A H W 1970 in *Current Topics in Radiation Research* **6** 71
Friedell H L, Thomas C and Krohmer J 1951 *Am. J. Roentgenol.* **65** 232
—— 1954 *Am. J. Roentgenol.* **71** 25
Gantchew M G and Rosenwald J C 1976 *Phys. Med. Biol.* **21** 209
Germain J St 1975 in *Handbook of Interstitial Brachytherapy* ed B S Hilaris (Acton, MA: Publishing Sciences Group) Ch 7 p 117
Gibbs R and Massey J B 1980 *Br. J. Radiol.* **53** 1100
Gillin M T, Kline R F, Wilson J F, and Cox J D 1984 *Int. J. Radiat. Oncol. Biol. Phys.* **10** 921
Glasgow G P 1981 in Recent Advances in Brachytherapy Physics *Medical Physics Monograph No* 7 ed D R Shearer (New York: American Institute of Physics) p 104
Glasgow G P and Dillman L T 1979 *Med. Phys.* **6** 49
—— 1982 *Med. Phys.* **9** 250
Glasgow G P, Walker S and Williams T D 1985 *Health Phys.* **48** 97
Glasser O, Quimby E H, Taylor L S, Weatherwax J L and Morgan R H 1961 *Physical Foundations of Radiology* 3rd edn (New York: Harper (Hoeber))
Glazebrook G A 1974 *Am. J. Roentgenol.* **120** 88
Goitein M 1976 *Clin. Radiol.* **27** 389
Green A E Jr, Broadwater J R and Hancock J A Jr 1969 *Am. J. Roentgenol.* **105** 609
Greening J R 1981 *Fundamentals of Radiation Dosimetry* (Bristol: Adam Hilger)
—— 1985 *Fundamentals of Radiation Dosimetry* 2nd edn (Bristol: Adam Hilger)
Hale J 1958 *Am. J. Roentgenol.* **79** 49
Hall E J 1972 *Br. J. Radiol.* **45** 81
Hall E J and Bedford J S 1964 *Radiat. Res.* **22** 305
Hall E J, Bedford J S and Oliver R 1966a *Br. J. Radiol.* **39** 302
Hall E J, Oliver R and Shepstone B J 1966b *Acta Radiol.* **14** 155
Hall E J and Rossi H 1975 *Br. J. Radiol.* **48** 477
Hames F 1937 *Am. J. Surg.* **38** 235
Haybittle J L 1955 *Br. J. Radiol.* **28** 320
—— 1957 *Br. J. Radiol.* **30** 49
Haybittle J L and Mitchell J H 1975 *Br. J. Radiol.* **48** 295
Henschke U K 1956 in *Therapeutic Use of Artificial Radioisotopes* ed P F Hahn (New York: John Wiley) p 375
—— 1958 in *The Treatment of Cancer and Allied Diseases* ed G T Pack and I M Ariel (New York: Hoeber)
—— 1960 *Radiology* **74** 834
Henschke U K and Cevc P 1968 *Radiobiol. Radiother.* **9** 287
Henschke U K, Hilaris B S and Mahan G D 1963 *Am. J. Roentgenol.* **90** 386
Henschke U K, James A G and Myers W G 1953 *Nucleonics* **11** 46
Herz R H 1969 *The Photographic Action of Ionising Radiations* (New York: Wiley-Interscience)
Heyman J 1929 *Acta Radiol.* **10** 49
—— 1935 *Acta Radiol.* **16** 129

Heyman J, Reuterwall O and Benner S 1941 *Acta Radiol.* **22** 14
Hidalgo J U, Spear V D, Garcia M, Madvell C R and Burke R 1967 *Am. J. Roentgenol.* **100** 852
Hilaris B S 1975a *Proc. 2nd Int. Symp. on Radiation Therapy (New York)* (New York: Memorial Sloan-Kettering Cancer Center)
—— 1975b *Handbook of Interstitial Brachytherapy* ed B S Hilaris (Acton, MA: Publishing Sciences Group)
Hilaris B S, Holt J G and Germain J St 1975 in *The Use of Iodine-125 Interstitial Therapy* DHEW Publication 76–8022 (Rockville, Md: US Department of Health, Education and Welfare) p 3
Hilaris B S, Kim J H and Tokita N 1976 *Am. J. Roentgenol.* **126** 171
Hilaris B S, Whitmore W F, Batata M A and Barzell W 1977 *Int. J. Radiat. Oncol. Biol. Phys.* **2** 631
Hilaris B S, Whitmore W F, Batata M A and Grabstald H 1974 *Am. J. Roentgenol.* **121** 832
Hine G J and Freidman M 1950 *Am. J. Roentgenol. Radium Ther.* **64** 989
Hodt H J, Sinclair W K and Smithers D W 1952 *Br. J. Radiol.* **25** 419
Holt J G 1956 *Am. J. Roentgenol.* **75** 476
Hope C S, Laurie J, Orr J S and Walters J H 1964 *Phys. Med. Biol.* **9** 345
Hope-Stone H F, Klevenhagen S C, Mantell B S, Morgan W Y and Scholnick S A 1981 *Clin. Radiol.* **32** 17
Horowitz Y S 1981 *Phys. Med. Biol.* **26** 765
Horsler A F C, Jones J C and Stacey A J 1964 *Br. J. Radiol.* **37** 385
Horwitz H, Kereiakes J G, Bahr G K, Cluxton S E and Barrett C M 1964 *Am. J. Roentgenol.* **91** 176
Howells R and Oliver R 1964 *Br. J. Radiol.* **37** 844
Hubbell J H 1969 *Report NSRDS-NBS-29* (Washington, DC: US Department of Commerce)
Hughes H A 1956 *Br. J. Radiol.* **19** 116
Hurdon E 1941 *Am. J. Roentgenol.* **45** 250
Inoue T, Hori S, Miyata Y, Shigematsu Y, Fuchihata H and Tanaka Y 1978 *Acta Radiol. Oncol.* **17** 27
Innes G S 1962 *Proc. Symp. on Ocular and Adnexal Tumours* (St Louis, Mo: Mosby)
International Atomic Energy Agency (IAEA) 1967 *Report Series No* 75 (Vienna: International Atomic Energy Agency)
International Commission on Radiological Protection (ICRP) 1977 Publ. 26 Recommendations of the Int. Commission on Radiological Protection *Ann. ICRP* **1** No 3
—— 1978 Statement from the 1978 Stockholm Meeting of the Int. Commission on Radiological Protection *Ann. ICRP* **2** No 1
International Commission on Radiological Units and Measurements (ICRU) 1951 *Br. J. Radiol.* **24** 54
—— 1954 *Br. J. Radiol.* **27** 243
—— 1957 *Report No* 8 (Washington, DC: ICRU Publications)
—— 1961 *Report No* 9 (Washington, DC: ICRU Publications)
—— 1962 *Radiation Quantities and Units Report No* 10a (Washington, DC: ICRU Publications)
—— 1968 *Radiation Quantities and Units Report No* 11 (Washington, DC: ICRU Publications)

—— 1970 *Specification of High Activity Gamma Ray Sources Report No* 18 (Washington, DC: ICRU Publications)
—— 1971 *Radiation Quantities and Units Report No* 19 (Washington, DC: ICRU Publications)
—— 1978 *Dose Specification for Reporting External Beam Therapy with Photons and Electrons Report No* 29 (Washington, DC: ICRU Publications)
—— 1979 *Average Energy Required to Produce an Ion Pair Report No* 31 (Washington DC: ICRU Publications)
—— 1980 *Radiation Quantities and Units Report No* 33 (Washington, DC: ICRU Publications)
—— 1985 *Dose and Volume Specification for Reporting Intracavitary Therapy in Gynaecology Report No* 38 (Washington, DC: ICRU Publications)
Jackson A W and Davies M L 1983 in *Radiation Treatment Planning* ed N M Bleehan, E Glastien and J L Haybittle (New York: Marcel Dekker) Ch 16
Jameson D G 1975 *Br. J. Radiol.* **48** 827
Jameson D G, Campion J M and Trevelyan A 1968 *Br. J. Radiol.* **41** 696
Jameson D G and Trevelyan A 1969 *Br. J. Radiol.* **42** 57
Jayaraman S and Agarwal S K 1982 *Strahlentherapie* **158** 107
Jayaraman S, Lanzl L H and Agarwal S K 1983 *Med. Phys.* **10** 871
Joelsson I and Bäckström A 1969 *Acta Radiol.* **8** 343
Johns H E and Cunningham J 1983 *The Physics of Radiology* 4th edn (Springfield, Ill: Charles C Thomas)
Johnsson J E and Nordberg U-B 1975 *Acta Radiol.* **14** 251
Jones C H and Dermentzoglou F 1971 *Br. J. Radiol.* **44** 203
Jones C H, Taylor K W and Stedeford J B H 1965 *Br. J. Radiol.* **38** 622
Jones E, Mallard J R and El Manharawy M S 1963 *Phys. Med. Biol.* **8** 59
Jones J C, Milan S and Lillicrap S C 1972 *Br. J. Radiol.* **45** 684
Jones J C and Stacey A J 1965 *Br. J. Radiol.* **38** 870
Joslin C A F and Smith C W 1970 *Proc. R. Soc. Med.* **63** 1029
Joslin C A F, Smith C W and Mallik A 1972 *Br. J. Radiol.* **45** 257
Kartha K I P, Kenny G N and Cameron J R 1966 *Am. J. Roentgenol.* **96** 66
Kemp L A W and Hall S M 1952 *Br. J. Radiol.* **25** 339
Keyser G M 1951 *Can. J. Phys.* **29** 301
Kim J H and Hilaris B S 1975 *Am. J. Roentgenol.* **123** 163
Kim Y S 1976 *Radiology* **120** 413
Kirk J, Gray W M and Watson E R 1971 *Clin. Radiol.* **22** 145
—— 1972 *Clin. Radiol.* **23** 93
—— 1973 *Clin. Radiol.* **24** 1
—— 1975a *Clin. Radiol.* **26** 77
—— 1975b *Clin. Radiol.* **26** 159
—— 1977a *Clin. Radiol.* **28** 29
—— 1977b *Clin. Radiol.* **28** 75
Klevenhagen S C 1973 *Br. J. Radiol.* **46** 1073
—— 1976 *PhD Thesis* University of London
—— 1978 *Med. Phys.* **5** 52
Kornelsen R O and Young M E J 1981 *Br. J. Radiol.* **54** 136
Kottmeier H L 1951 *J. Fac. Radiol.* **2** 321
—— 1959 *Am. J. Obstet. Gynaecol.* **78** 1127

—— 1964 *Acta Obstet. Gynaecol. Scand.* **43** Suppl. 2 1
Krishnaswamy V 1971 *Radiology* **98** 155
—— 1972a *Radiology* **105** 181
—— 1972b *Phys. Med. Biol.* **17** 56
—— 1974 *Phys. Med. Biol.* **19** 886
—— 1978 *Radiology* **126** 489
—— 1979 *Radiology* **132** 727
Kuan H M and Anderson L L 1977 *Med. Phys.* **4** 357
Kubiatowicz D O 1981 in Recent Advances in Brachytherapy Physics *Medical Physics Monograph No* 7 ed D R Shearer (New York: American Institute of Physics) p 57
Kwan D K, Kagan A R, Olch A J, Chan P Y M, Hintz B L and Wollin M 1983 *Med. Phys.* **10** 456
Lathja L G and Oliver R 1961 *Br. J. Radiol.* **34** 252
Laughlin J S, Siler W M. Holodny E I and Ritter E W 1963 *Am. J. Roentgenol.* **89** 470
Lewis G C, Raventos A and Hale J 1960 *Am. J. Roentgenol.* **83** 432
Lin F-M and Cameron J R 1967 *Am. J. Roentgenol.* **100** 863
Ling C C, Anderson L L and Biggs P J 1981 in Recent Advances in Brachytherapy Physics *Medical Physics Monograph No* 7 ed D R Shearer (New York: American Institute of Physics) p 115
Ling C C, Anderson L L and Shipley W V 1979 *Int. J. Radiat. Oncol. Biol. Phys.* **5** 419
Ling C C, Gromadzki Z C, Rustgi S N and Cundiff J H 1983a *Radiology* **146** 791
Ling C C, Yorke E D, Spiro I J, Kubiatowicz D and Bennett D 1983b *Int. J. Radiat. Oncol. Biol. Phys.* **9** 1747
Liu J and Edwards F M 1979 *Radiology* **132** 748
Liversage W E 1969 *Br. J. Radiol.* **42** 432
Liversage W E, Martin-Smith P and Ramsey N W 1967 *Br. J. Radiol.* **40** 887
Loevinger R 1953 *Rev. Sci. Instrum.* **24** 907
—— 1981 in Recent Advances in Brachytherapy Physics *Medical Physics Monograph No* 7 ed D R Shearer (New York: American Institute of Physics) p 22
Loevinger R, Japha E M and Brownell G L 1956 in *Radiation Dosimetry* ed G J Hine and G L Brownell (New York: Academic) Ch 16
Loevinger R and Trott N 1966 *Int. J. Appl. Radiat. Isotopes* **17** 103
McKinlay A F 1981 *Thermoluminescence Dosimetry* (Bristol: Adam Hilger)
Marinello G, Dutreix A, Pierquin B and Chassagne D 1978 *J. Radiol. Electrol.* **59** 621
Marinello G, Valero M, Leung S and Pierquin B 1985 *Int. J. Radiol. Oncol. Biol. Phys.* **11** 1973
Maruyama Y 1979 *Radiology* **133** 473
Maruyama Y, Beach J L and Feola J M 1980a *Radiology* **137** 775
Maruyama Y, Feola J M and Beach J L 1983 *Int. J. Radiat. Oncol. Biol. Phys.* **9** 1715
Maruyama Y, Van Nagell J R Jr, Wrede D E, Coffey C W III, Utley J R and Avila J 1976 *Radiology* **120** 389
Maruyama Y, Yoneda J, Krolikiewicz H, Mendiondo O, Beach J L, Coffey C W III, Thompson D, Wilson L C, Feola J M, Van Nagell J R, Donaldson E S and Powell D 1980b *Int. J. Radiat. Oncol. Biol. Phys.* **6** 1629
Massey J B, Pointon R C S and Wilkinson J M 1985 *Br. J. Radiol.* **58** 911
Masterson M E, Simpson L D and Mohan R 1979 *Med. Phys.* **6** 347
Mayles W P M, Mayles H M O and Turner P C 1985 *Br. J. Radiol.* **58** 529
Mayneord W V 1945 *Br. J. Radiol.* **18** 12

Meertens H 1985 in *Brachytherapy 1984* ed R F Mould (Holland: Nucletron Trading B V)
Meisberger L L, Keller R J and Shalek R J 1968 *Radiology* **90** 953
Meredith W J 1947 *Radium Dosage, The Manchester System* ed W J Meredith (Edinburgh and London: E and S Livingstone)
—— 1967 *Radium Dosage, The Manchester System* ed W J Meredith 2nd edn (Edinburgh and London: E and S Livingstone)
Meredith W J, Greene D and Kawashima K 1966 *Br. J. Radiol.* **39** 280
Meredith W J and Massey J B 1977 *Fundamental Physics of Radiology* 3rd edn (Bristol: John Wright)
Meredith W J and Stephenson S K 1967 in *Radium Dosage, The Manchester System* ed W J Meredith 2nd edn (Edinburgh and London: E and S Livingstone)
Meurk M L, Jacobson A and Schultz R J 1957 *Radiology* **68** 256
Milan S 1975 *MSc Thesis* University of London
Mohan R 1981 in Recent Advances in Brachytherapy Physics *Medical Physics Monograph No* 7 ed D R Shearer (New York: American Institute of Physics) p 134
Mohan R and Anderson L L 1974 in *BRACHY Interstitial and Intracavitary Dose Computation Program Users Guide* (New York: Memorial Sloan-Kettering Cancer Center)
—— 1978 in *BRACHY II Interstitial and Intracavitary Dose Computation Program Users Guide* (New York: Memorial Sloan-Kettering Cancer Center)
—— 1981 in *BRACHY II Interstitial and Intracavitary Dose Computation Program Users Guide* Section II 2nd edn (New York: Memorial Sloan-Kettering Cancer Center)
Morphis O L 1960 *Am. J. Roentgenol. Radiat. Ther. Nucl. Med.* **83** 455
Morris A H 1964 *MD Thesis* Yale University, Connecticut
Morton J L, Callendine G W Jr and Myers W G 1951 *Radiology* **56** 553
Mould R F 1966 *MSc Thesis* University of London
—— 1980 in *A History of X-rays and Radium* (London: IPC Business Press)
Mould R F and Hobday P A 1984 *Radiation Dosimetry for the Amersham Afterloading System for Gynaecological Brachytherapy* (London: Amersham International plc)
Murphy W T 1967 *Radiation Therapy* (Philadelphia and London: Saunders)
Mussel L E 1956a *Br. J. Radiol.* **29** 402
—— 1956b *Am. J. Roentgenol.* **75** 497
Myers W G 1948 *Am. J. Roentgenol.* **60** 816
Myers W G, Colmeny B H and McLellon W M 1953 *Am. J. Roentgenol.* **70** 258
National Bureau of Standards (NBS) 1982 *NBS Handbook No* 138 ed T N Padikal (Washington, DC: US Government Printing Office)
National Commission on Radiological Protection (NCRP) 1974 *Specification of Gamma-ray Brachytherapy Sources Report No* 41 (Washington, DC: NCRP Publications)
—— 1976 *Structural Shielding Design and Evaluation for Medical Use of X-rays and Gamma-rays of Energies up to 10 MeV Report No* 49 (Washington, DC: NCRP Publications)
Neary G J 1942 *Br. J. Radiol.* **15** 104
—— 1943 *Br. J. Radiol.* **16** 225
Nelson R F and Meurk M L 1958 *Radiology* **70** 90
Nuttall J R and Spiers F W 1946 *Br. J. Radiol.* **19** 135
O'Connell D, Howard N, Joslin C A F, Ramsey N W and Liversage W E 1965 *Lancet* **2** 570
O'Connell D, Joslin C A F, Howard N, Ramsey N W and Liversage W E 1967 *Br. J. Radiol.* **40** 882
Oddie T H 1940 *Br. J. Radiol.* **13** 389

Onai Y, Tomaru T, Irifuna T, Uchida I I, Tsuya T and Kaneta K 1978 *Nippon Acta Radiol.* **38** 642
Orton C G 1974 *Br. J. Radiol.* **47** 603
—— 1980 in *Handbook of Medical Physics* vol 1 ed R G Waggener, J G Kereiakes and R J Shalek (Florida: CRC)
Orton C G and Ellis F 1973 *Br. J. Radiol.* **46** 529
Paine C H 1972 *Clin. Radiol.* **23** 263
Parker H M 1947 in *Radium Dosage, The Manchester System* ed W J Meredith (Edinburgh and London: E and S Livingstone) Ch 7, 8, 10
Paterson R 1963 *The Treatment of Malignant Disease by Radiotherapy* 2nd edn (London: E Arnold)
Paterson R and Parker H M 1934 *Br. J. Radiol.* **7** 592
—— 1938 *Br. J. Radiol.* **11** 252, 313
Paterson R, Parker H M and Spiers F W 1936 *Br. J. Radiol.* **9** 487
Payne W H and Waggener R G 1974a *Med. Phys.* **1** 165
—— 1974b *Med. Phys.* **1** 210
Permar P H 1976 *Int. J. Radiat. Oncol. Biol. Phys.* **1** 1003
Pierquin B 1964 *Precis de Curietherapie* (Paris: Masson)
—— 1966 *Radiol. Clin. Biol.* **35** 407
—— 1971 *Proc. Conf. on Afterloading in Radiotherapy (New York) May 1971* DHEW Publication 72–8024 (Rockville, Md: US Department of Health, Education and Welfare, Bureau of Radiological Health) p 204
Pierquin B and Chassagne D J 1962 *J. Radiol.* **43** 65
Pierquin B, Chassagne D, Baillet F and Paine C H 1973 *Clin. Radiol.* **24** 506
Pierquin B, Chassagne D J and Cox J D 1971 *Radiology* **99** 661
Pierquin B, Chassagne D, Chahbazian C H and Wilson J F 1978a *Brachytherapy* (St Louis, Mo: Warren H Green)
Pierquin B and Dutreix A 1966 *Ann. Radiol.* **9** 757
—— 1967 *Br. J. Radiol.* **40** 184
Pierquin B, Dutreix A, Paine C H, Chassagne D, Marinello G and Ash D 1978b *Acta Radiol.* **17** 33
Pierquin B, Dutreix A, Wambersie A and Chassagne D 1969 *J. Radiol. Electrol.* **50** 377
Pierquin B and Fayos J V 1962 *Am. J. Roentgenol.* **87** 585
Planskoy B 1980 *Phys. Med. Biol.* **25** 519
—— 1983 in *Radiation Treatment Planning* ed N M Bleehan, E Glastien and J L Haybittle (New York: Marcel Dekker) Ch 3
Porter E H 1970 *Br. J. Radiol* **43** 629
Potish R A, Deibel F C and Khan F M 1982 *Radiology* **145** 479
Powers W E, Bogardus C R, White W and Gallagher T 1965 *Radiology* **85** 135
Quimby E H 1922 *Am. J. Roentgenol.* **9** 671
—— 1932 *Am. J. Roentgenol.* **27** 18
—— 1944 *Radiology* **43** 572
Quimby E H and Castro V 1953 *Am. J. Roentgenol.* **70** 739
Ramm W J 1966 in *Radiation Dosimetry* vol II ed F H Attix and W C Roesch (New York: Academic) p 126
Rao G U V, Kan P T and Howells R 1981 *Int. J. Radiat. Oncol. Biol. Phys.* **7** 431
Redpath A T, Douglas M A and Orr J A 1976 *Br. J. Radiol.* **49** 963
Regaud C 1929 *Am. J. Roentgenol.* **21** 1

Reuss A and Brunner F 1957 *Strahlentherapie* **103** 279
Richman M, Memula H and Doss L L 1980 *Br. J. Radiol.* **53** 996
Ridings G R 1963 *Am. J. Roentgenol.* **89** 500
Rose J A, Bloedorn F G and Robinson J E 1966 *Am. J. Roentgenol.* **97** 1032
Rose J A, Robinson J E and Bloedorn F G 1967 *Am. J. Roentgenol.* **100** 878
Rosemark P J, Tobochnik N, Herman M W, Whiting J S, Thompson R W, Weisenburger T H, Metcalf D R and Greenfield M A 1982 *Radiology* **142** 517
Rosen I I, Khan K M, Lane R G and Kelsey G A 1982 *Med. Phys.* **9** 220
Rosenthal M S and Nath R 1983 *Med. Phys.* **10** 475
Rosenwald J C 1970 *Ann. Phys. Biol. Med.* **3** 139
—— 1975 *Comput. Programs Biomed.* **4** 103
Rubin P, Gerle R D, Quick R S and Greenlaw R H 1963 *Am. J. Roentgenol.* **89** 91
Santhamma A V and Das K R 1978 *Br. J. Radiol.* **51** 507
Saylor W L and Dillard M 1976 *Med. Phys.* **3** 17
Schlea C S and Stoddard D H 1965 *Nature* **206** 1058
Schlienger M, Rosenwald J C, Miculutia M, Quint R and Pierquin B 1970 *Acta Radiol.* **9** 282
Schulz R J, Chandra R and Nath R 1980 *Med. Phys.* **7** 355
Schwarz G 1969 *Am. J. Roentgenol.* **105** 579
Scott W P 1972a *Am. J. Roentgenol.* **119** 620
—— 1972b *Radiology* **105** 454
—— 1977 *Radiology* **122** 832
Seydel G 1979 in Californium-252 Radiobiology and Medical Applications *Int. Symp. on Californium-252 Utilisation (Brussels) 1976* vol 1 ed D J Mewissen (SC de Pont de Nemurs)
Shalek R J and Stovall M 1961 *Radiology* **76** 119
—— 1962 in *Radiation Therapy in the Management of Cancer of the Oral Cavity and Oropharynx* ed G H Fletcher and W S MacCabe (Springfield, Ill: C C Thomas) p 293
—— 1968 *Am. J. Roentgenol. Radium Ther.* **102** 662
—— 1969 in *Radiation Dosimetry* vol III ed F H Attix and W C Roesch (New York: Academic) Ch 31 p 743
Shalek R J, Stovall M and Sampiere V A 1957 *Am. J. Roentgenol.* **77** 863
Sharma S C, Bello J E and Arbath F G 1979 *Med. Phys.* **6** 157
Sharma S C, Williamson J F and Cytacki E 1982 *Int. J. Radiat. Oncol. Biol. Phys.* **8** 1803
Shapiro A, Schwartz B, Windham J P and Kereiakes J G 1976 *Med. Phys.* **3** 241
Shaw J E and Thomas R L 1973 *Br. J. Radiol.* **47** 634
Sheldon J J, Vuksanovic M, Wold G J and Fonts E A 1969 *Am. J. Roentgenol.* **105** 120
Sievert R 1921 *Acta Radiol.* **1** 89
—— 1932 *Acta Radiol. Suppl.* **14**
Sinclair W K 1952 *Br. J. Radiol.* **25** 417
Sinclair W K and Trott N G 1956 *Br. J. Radiol.* **29** 15
Smocovitis D, Young M E J and Batho H F 1967 *Br. J. Radiol.* **40** 771
Snelling M, Ellis R E and Jameson D G 1965 in *Progress in Radiology* vol III (Rome: International Congress of Radiology) p 1709
Souttar H S 1931 *Br. J. Radiol.* **4** 681
Spiers F W 1947 in *Radium Dosage, The Manchester System* ed W J Meredith (Edinburgh and London: E and S Livingstone) Ch 9
Spira J and Horn R 1967 *Am. J. Roentgenol.* **100** 858

Stallard H B 1962 *Trans. Ophthalmol. Soc. UK* **82** 473
Stephens S O 1981 in Recent Advances in Brachytherapy Physics *Medical Physics Monograph No* 7 ed D R Shearer (New York: American Institute of Physics) p 72
Storchi P R M and Van Kleffens H J 1979 *Comput. Programs Biomed.* **9** 141
Storm E and Israel H L 1970 in *Nuclear Data Tables* A7 (New York and London: Academic) p 565
Stovall M and Shalek R J 1962 *Radiology* **78** 950
—— 1968 *Am. J. Roentgenol* **102** 677
—— 1972 *Comput. Programs Biomed.* **2** 125
Strickland P 1971 in *Gynaecological Cancer* ed T J Deeley (London: Butterworths) Ch 11
Suit H D, Moore E B, Fletcher G H and Worsnop R 1963 *Radiology* **81** 126
Suit H D, Shalek R J, Moore E B and Andrews J R 1961 *Radiology* **76** 431
Suthanthuran K 1981 in Recent Advances in Brachytherapy Physics *Medical Physics Monograph No* 7 ed D R Shearer (New York: American Institute of Physics) p 49
Tai D L and Maruyama Y 1978 *Radiology* **128** 795
Ter-Pogossian M, Ittner W B III and Aly S M 1952 *Nucleonics* **10** No 6 50
Tochilin E 1955 *Am. J. Roentgenol.* **73** 265
Tod M C and Meredith W J 1938 *Br. J. Radiol.* **11** 809
—— 1953 *Br. J. Radiol.* **26** 252
Todd T F 1938 *Surg. Gynaecol. Obstet.* **67** 617
Toepfer K-D and Rosenow U 1980 *Br. J. Radiol.* **53** 1078
Tokita N, Kim J H and Hilaris B 1980 *Int. J. Radiat. Oncol. Biol. Phys.* **6** 1745
Tranter F W 1959 *Br. J. Radiol.* **32** 350
Tripathi U B and Shanta A 1985 *Med. Phys.* **12** 88
Trucco E 1962 Biological and Research Division, Argonne National Laboratory *Semi-annual Report* ANL 6723 (Chicago: Argonne National Laboratory)
Tsuya A, Kaneta K, Sugiyama T, Onai Y, Irifune T, Tomaru T, Uchida M, Komata S and Tsuchida Y 1979 *Nippon Acta Radiol.* **39** 370
Tudway R C 1953 *Acta Radiol.* **39** 415
Turner D 1909 Report of meeting *Lancet* **2** 1873
Twiss D H and Bradshaw A L 1970 *Br. J. Radiol.* **43** 48
Vaeth J M and Meurk M L 1963 *Am. J. Roentgenol.* **89** 87
Vallejo A, Hilaris B S and Anderson L L 1977 *Int. J. Radiat. Oncol. Biol. Phys.* **2** 731
Van Dilla M A and Hine G J 1952 *Nucleonics* **10** No 7 54
Van Kleffens H J and Star W M 1979 *Int. J. Radiat. Oncol. Biol. Phys.* **5** 557
Vickery B L and Redpath A T 1977 *Br. J. Radiol.* **50** 280
Walstram R 1965 *Acta Radiol. Suppl.* **236** 84
Wang C C 1975 *Radiology* **117** 225
Waterman F M and Strubler K A 1983 *Med. Phys.* **10** 155
Webb S and Fox R A 1979 *Br. J. Radiol.* **52** 482
Welsh A D, Dixon-Brown A and Stedeford J B H 1983 *Acta Radiol.* **22** 331
Whelpton D and Watson B J 1963 *Phys. Med. Biol.* **8** 33
Whyte G N 1955 *Br. J. Radiol.* **28** 635
Wickham L and Degrais M 1910 *Radiumtherapy* (London: Cassell)
Wiernik G and Young C M A 1979 in *High LET Radiation in Clinical Radiotherapy* ed G W Barendson, J J Broerse and K Breuer (Oxford: Pergamon) p 206
Wilkinson J M 1972 *Br. J. Radiol.* **45** 708
Wilkinson J M, Hendry J H and Hunter R D 1980 *Br. J. Radiol.* **53** 890

Wilkinson J M, Moore C J, Notley H M and Hunter R D 1983 *Br. J. Radiol.* **56** 409
Williamson J F, Khan F M, Sharma S C and Fullerton G D 1982 *Radiology* **142** 511
Williamson J F, Morin R L and Khan F M 1983a *Phys. Med. Biol.* **28** 1021
—— 1983b *Med. Phys.* **10** 135
Wood R G 1981 *Computers in Radiotherapy Planning* (Chichester: John Wiley)
Woods M J 1970 *Int. J. Appl. Radiat. Isotopes* **21** 752
Woods M J, Callow W J and Christmas P 1983 *Int. J. Nucl. Med. Biol.* **10** 127
Woods M J and Lucas S E M 1974 *Int. J. Appl. Radiat. Isotopes* **26** 488
Wootton P, Shalek R J and Fletcher G H 1954 *Am. J. Roentgenol.* **71** 683
Wu A, Zwicker R D and Sternick E S 1985 *Med. Phys.* **12** 27
Wyckoff H O 1983 *Med. Phys.* **10** 715
Young M E J and Batho H F 1964 *Br. J. Radiol* **37** 38

Index